Determination of Human-Health Pharmaceuticals in Filtered Water by Chemically Modified Styrene-Divinylbenzene Resin-Based Solid-Phase Extraction and High-Performance Liquid Chromatography/Mass Spectrometry

By Edward T. Furlong, Stephen L. Werner, Bruce D. Anderson, and Jeffery D. Cahill

Chapter 5
Section B, Methods of the National Water Quality Laboratory
Book 5, Laboratory Analysis

Techniques and Methods 5–B5

U.S. Department of the Interior
U.S. Geological Survey

U.S. Department of the Interior
DIRK KEMPTHORNE, Secretary

U.S. Geological Survey
Mark D. Myers, Director

U.S. Geological Survey, Reston, Virginia: 2008

For product and ordering information:
World Wide Web: http://www.usgs.gov/pubprod
Telephone: 1-888-ASK-USGS

For more information on the USGS—the Federal source for science about the Earth, its natural and living resources, natural hazards, and the environment:
World Wide Web: http://www.usgs.gov
Telephone: 1-888-ASK-USGS

Suggested citation:
Furlong, E.T., Werner, S.L., Anderson, B.D., and Cahill, J.D., 2008, Determination of human-health pharmaceuticals in filtered water by chemically modified styrene-divinylbenzene resin-based solid-phase extraction and high-performance liquid chromatography/mass spectrometry: U.S. Geological Survey Techniques and Methods, book 5, sec. B, chap. B5, 56 p.

Contents

Figures

Tables

Conversion Factors

Multiply	By	To obtain
	Length	
centimeter (cm)	0.3937	inch
micrometer (μm)	3.937×10^{-5}	inch
millimeter (mm)	0.03937	inch
meter (m)	3.281	foot
	Volume	
liter (L)	33.82	ounce, fluid
milliliter (mL)	0.0338	ounce, fluid
microliter (μL)	0.0338×10^{-3}	ounce, fluid
	Mass	
gram (g)	0.03527	ounce, avoirdupois
milligram (mg)	3.53×10^{-5}	ounce
microgram (μg)	3.53×10^{-8}	ounce
nanogram (ng)	3.53×10^{-11}	ounce
	Pressure	
kilopascal (kPa)	0.1450	pound per square inch
millitorr (mtorr)	1.93×10^{-5}	pound per square inch
	Concentration, in water	
nanograms per liter (ng/L)	1	parts per trillion
micrograms per liter (μg/L)	1	parts per billion
milligrams per liter (mg/L)	1	parts per million

Temperature in degrees Celsius (°C) may be converted to degrees Fahrenheit (°F) as follows:

$$°F=(1.8\times°C)+32$$

Abbreviated water-quality units

g/L	gram per liter
g/mL	gram per milliliter
kg/m²	kilogram per square meter
μg/L	microgram per liter
μg/μL	microgram per microliter
μg/mL	microgram per milliliter
μL/L	microliter per liter
μL/mL	microliter per milliliter
mg/L	milligram per liter
mg/mL	milligram per milliliter
mL/min	milliliter per minute
mM	millimolar
ng/L	nanogram per liter
ng/μg	nanogram per microgram
ng/μL	nanogram per microliter
ng/mL	nanogram per milliliter

Abbreviations and acronyms

amu	atomic mass unit
APCI	atmospheric pressure chemical ionization
cat. no.	catalog number
CCB	continuing calibration blank
CCV	continuing calibration verification
DOC	dissolved organic carbon
ESI	electrospray ionization
FEB	field equipment blank
GC	gas chromatography
GC/MS	gas chromatography/mass spectrometry
GMD	grand median
GMN	grand mean
GSD	grand standard deviation
HPLC	high-performance liquid chromatography
HPLC/MS	high-performance liquid chromatography/mass spectrometry
HPLC/MSD	high-performance liquid chromatography/mass selective detector
IRL	interim reporting level
kV	kilovolt
LC	liquid chromatography
LC/MS	liquid chromatography/mass spectrometry
LC/MSD	liquid chromatography/mass selective detector
LMS	laboratory matrix spike
LOQ	limit of quantitation
LRB	laboratory reagent blank
LRL	laboratory reporting level
LRS	laboratory reagent spike
LT–MDL	long-term method detection level
M	molarity (moles per liter)
mbar	millibar
mL	milliliter
mM	millimolar
MDL	method detection limit
MHz	megahertz
MS	mass spectrometry
MSD	mass selective detector
m/z	mass-to-charge ratio
μm	micrometer
NAWQA	National Water-Quality Assessment Program
NWIS	National Water Information System
NWQL	National Water Quality Laboratory

Abbreviations and acronyms—Continued

ODS	octadecylsilane
POC	polar organic compound
PTFE	polytetrafluoroethylene
QA/QC	quality assurance/quality control
QC	quality control
RF	response factor
SPE	solid-phase extraction
TFA	trifluoroacetic acid
™	trademark
USEPA	U.S. Environmental Protection Agency
USGS	U.S. Geological Survey
UV	ultraviolet
v/v	volume per volume
w/v	weight per volume
$\cong$	equals about
$\pm$	plus or minus
$\equiv$	identical with, congruent
>	greater than
<	less than
~	approximately
®	registered trademark

Determination of Human-Health Pharmaceuticals in Filtered Water by Chemically Modified Styrene-Divinylbenzene Resin-Based Solid-Phase Extraction and High-Performance Liquid Chromatography/Mass Spectrometry

By Edward T. Furlong, Stephen L. Werner, Bruce D. Anderson, and Jeffery D. Cahill

Abstract

In 1999, the Methods Research and Development Program of the U.S. Geological Survey National Water Quality Laboratory began the process of developing a method designed to identify and quantify human-health pharmaceuticals in four filtered water-sample types: reagent water, ground water, surface water minimally affected by human contributions, and surface water that contains a substantial fraction of treated wastewater. Compounds derived from human pharmaceutical and personal-care product use, which enter the environment through wastewater discharge, are a newly emerging area of concern; this method was intended to fulfill the need for a highly sensitive and highly selective means to identify and quantify 14 commonly used human pharmaceuticals in filtered-water samples. The concentrations of 12 pharmaceuticals are reported without qualification; the concentrations of two pharmaceuticals are reported as estimates because long-term reagent-spike sample recoveries fall below acceptance criteria for reporting concentrations without qualification.

The method uses a chemically modified styrene-divinylbenzene resin-based solid-phase extraction (SPE) cartridge for analyte isolation and concentration. For analyte detection and quantitation, an instrumental method was developed that used a high-performance liquid chromatography/mass spectrometry (HPLC/MS) system to separate the pharmaceuticals of interest from each other and coextracted material. Immediately following separation, the pharmaceuticals are ionized by electrospray ionization operated in the positive mode, and the positive ions produced are detected, identified, and quantified using a quadrupole mass spectrometer.

In this method, 1-liter water samples are first filtered, either in the field or in the laboratory, using a 0.7-micrometer (μm) nominal pore size glass-fiber filter to remove suspended solids. The filtered samples then are passed through cleaned and conditioned SPE cartridges at a rate of about 15 milliliters per minute. Excess water is eliminated from the cartridge sorbent bed by passing air through the cartridges, and the analytes retained on the SPE bed are eluted from the cartridge sequentially, first with methanol, followed by acidified methanol, and combined in collection tubes. This sample extract then is reduced from about 10 milliliters (mL) to about 0.1 mL (or 100 microliters) under a stream of purified nitrogen gas with the collection tubes in a heated (40°C) water bath. The reduced extracts then are fortified with an internal standard solution (when using internal standard quantitation), brought to a final volume of 1 mL with an aqueous ammonium formate buffer solution, and filtered through a 0.2-μm Teflon syringe filter as they are transferred into vials for instrumental analysis.

Instrumental analysis by the HPLC/MS procedure permits determination of individual pharmaceutical concentrations from 0.005 to 1.0 microgram per liter, based on the lowest and the highest calibration standards routinely used. The reporting levels for this method are compound dependent, and have been experimentally determined based on the precision of quantitation of compounds from eight fortified organic-free water samples in single-operator experiments. The method detection limits and interim reporting levels for the compounds determined by this method were calculated from recoveries of the pharmaceuticals from reagent-water samples amended at 0.05 microgram per liter, and ranged between 0.0069 and 0.0142 microgram per liter, and 0.015 and 0.10 microgram per liter, respectively. Concentrations for 12 compounds are reported without qualification, and for two compounds are reported as qualified estimates. After initial development, the method was applied to more than 1,800 surface-, ground-, and wastewater samples from 2002 to 2005 and documented in a number of published studies. This research application of the method provided the opportunity to collect a large data set of ambient environmental concentrations and also permitted the collection of an extensive set of reagent blanks and spike quality-control (QC) samples. This multiple-year set of QC samples enabled further evaluation of method performance under multiple operator and multiple instrument conditions typical of routine laboratory operation. These results are an important part of the entire data set contained in this report because they document method performance over an extended time. Because sample matrix can substantially affect method performance, inclusion of environmental matrix-spike samples is required as a routine component of study plan quality control.

Method performance has been measured by long-term tracking of observed recoveries from fortified organic-free water samples processed with environmental samples (laboratory reagent spikes), as well as by observed recoveries from multiple fortified environmental water samples. The fortified environmental samples included surface water, wastewater effluent-dominated surface water, and ground water, fortified at two environmentally relevant concentrations and corrected for ambient environmental concentrations.

Because the responses of individual pharmaceuticals vary as a function of proton affinity, the ionization efficiency, and thus relative response, of each pharmaceutical, the quality-control surrogate compounds, and the quantitation internal standard can be suppressed or enhanced by the presence of the sample matrix. As a result, several quality-control sample types are required to properly interpret the ambient environmental concentrations of pharmaceuticals in aqueous samples. The quality-control sample types and results include laboratory reagent spikes and laboratory reagent blanks to provide insight into the performance of the method in the absence of a sample matrix, and matrix-spike recovery samples and replicate environmental samples, collected from representative sample matrix types within the aquatic system under study.

Introduction

During the 1990s, it became widely recognized that pharmaceutical and other personal-care products used by humans presented a source of chemical contamination that had potential for human or ecosystem effects, and which had yet to be assessed (Daughton and Ternes, 1999). There was a dearth of reliable and sensitive analytical methods that could be used to test for a broad range of such potential contaminants, which were typically of multiple chemical classes. The Methods Research and Development Program of the U.S. Geological Survey National Water Quality Laboratory (NWQL) recognized a need for analytical methods for pharmaceuticals that would be rugged, reliable, amenable to a broad range of water types, and suitable for determining trace amounts of pharmaceuticals at the ambient environmental concentrations expected to be present in surface- and ground-water samples, typically less than a microgram per liter. Because human wastewater is an important source for these compounds, the method conceived would complement other official methods of the USGS that are used to measure anthropogenic waste indicators in water. These anthropogenic waste indicators represent a number of sources and uses, including ethoxylate surfactants, fragrances, food additives, antioxidants, phosphate flame retardants, plasticizers, industrial solvents, disinfectants, and fecal sterols (Zaugg and others, 2002).

Building on previous experience with the use of solid-phase extraction (SPE) and liquid chromatography/mass spectrometry (HPLC/MS) for the analysis of pesticides (Furlong and others, 2000, 2001) and literature reports describing the presence of pharmaceuticals in the environment, a multiclass method using SPE and HPLC/MS was developed (Cahill and others, 2004).

A critical step in the development of this method was the selection of chemically modified styrene-divinylbenzene resin-based SPE cartridges for multiclass pharmaceutical isolation. Cahill (2000) and Cahill and others (2004) showed that the Oasis® HLB (Waters Corporation) solid-phase sorbent was the optimal choice for pharmaceutical isolation from filtered water, due to overall superior recovery for all compounds, low coextractive interferences in the sorbent material, batch-to-batch comparability, resistance to polymeric structural breakdown, and competitive cost.

The pharmaceuticals included in this method were selected based on human use (as reflected by annual total prescriptions in the United States), typical active ingredient doses, likely persistence through human metabolism, and, after excretion, persistence through common wastewater-treatment processes. Moreover, the pharmaceuticals were experimentally determined to perform well during SPE isolation, and were shown to be efficiently ionized under the positive electrospray ionization conditions used in this method. Eighteen additional compounds were investigated for possible inclusion in this method, but were found to be either poorly recovered from water using SPE cartridges or not readily amenable to the electrospray HPLC/MS procedure used for analysis.

The method relies on positive mode ionization in the electrospray ionization interface (commonly referred to as the ion source). An additional ionization approach, known as atmospheric pressure chemical ionization (APCI), also was tested for use in the analytical procedure. However, the electrospray ionization technique proved to be best suited to the compounds selected for the method, likely because of their ability to adduct protons during the electrospray ionization process.

After initial development, the method was applied to more than 1,800 surface-, ground-, and wastewater samples from 2002 to 2005 and documented in a number of published studies. This research application of the method provided the opportunity to collect a large data set of ambient environmental concentrations. It is beyond the scope of this report to provide detailed analysis of these studies, but several articles (see Barber and others, 2006; Cahill and others, 2004; Glassmeyer and others, 2005; Kolpin and others, 2002; Sando and others, 2005; and Wilkison and others, 2006) provide useful information for evaluating how this method has been used to assess the presence and distribution of pharmaceuticals in aquatic environments, and how data produced using this method can be applied to understand water-quality questions. These research applications of the method also permitted the collection of an extensive set of laboratory reagent blank and spike quality-control (QC) samples. This multiple-year set of QC samples provided the opportunity to further evaluate method performance under multiple operator and multiple instrument conditions typical of routine laboratory operation. These results are an important part of the entire data set contained in this report because they document method performance over an extended time.

Analytical Method

Organic Compounds and Parameter Codes: Pharmaceuticals, dissolved, chemically modified styrene-divinylbenzene resin-based extraction, high-performance liquid chromatography/mass spectrometry, O-2080-08.

1. Scope and Application

This method is designed for the determination of human-use pharmaceuticals (table 1) in filtered water samples. The method is applicable to those compounds that (1) are efficiently partitioned from the water onto a chemically modified styrene-divinylbenzene resin-based solid-phase extraction (SPE) material, (2) can be quantitatively eluted from the SPE material, (3) can be reliably separated by liquid chromatography (LC), and (4) can be efficiently ionized by high-performance liquid chromatography/mass spectrometry (HPLC/MS) using an electrospray ionization (ESI) interface operated in the positive ionization mode. Because of the presence of matrix effects discussed later in this report, inclusion of environmental matrix-spike samples is required as a routine component of study plan quality control.

This method is applicable to filtered water samples, which in this case refers to natural-water samples that have been filtered with a pre-baked glass-fiber filter using the method of Wilde and others (2004). The performance characteristics of this method have been determined for a small set of surface- and ground-water samples, and users of this method need to recognize that performance characteristics of other matrices have not been tested. Any determinations made in new matrices would require appropriate qualification until an analogous performance evaluation had been made. Matrices, such as septage, wastewater influents or other liquids collected in wastewater-treatment facilities prior to discharge, and liquids collected from confined animal-feeding operations, among others, are known to contain complex coextracted interferences that compete for sorption on the SPE phase during elution, and in final sample extracts may suppress ionization of the pharmaceuticals of interest by competing for available charge during electrospray ionization (Enke, 1997). Thus, routine analyses of matrix-spike samples collected within the study area are required as a routine component of project quality control.

Two classes of determinations are reported for samples analyzed by this method. Compounds whose long-term recovery and variability fall within the criteria for acceptable performance and are reported without qualification [Furlong and others, 2001; NWQL Standard Operating Procedure MX0015.2, Guidelines for method validation and publication at the National Water Quality Laboratory (R.B. Green and W.T. Foreman, U.S. Geological Survey, written commun., 2005)]. Compounds for which long-term performance falls below the criteria for acceptable performance, but above the criteria for exclusion from the method, are reported as qualified estimates and indicated by a qualifier code of "E." The performance classification for each compound in the method is listed in table 1.

The same qualitative identification criteria are applied to all 14 pharmaceuticals, and all detections reported using this method must meet these qualitative identification criteria. Twelve pharmaceuticals are reported without qualification, and are reproducibly well-recovered using this method, as defined by median recoveries of an extended set (n=157) of laboratory reagent spike (LRS) samples between 60 and 120 percent, and by variation (as indicated by the nonparametric statistic f-pseudosigma) of less than 25 percent. Two pharmaceuticals are reported as qualified estimates because they do not meet these long-term method performance quantitation criteria, but are retained in the method because they are used in substantial quantity, have important environmental or toxicological effects, are appropriately qualitatively identified, and are recovered at a concentration of greater than 35 percent, coupled with a variance (as represented by the nonparametric statistic f-pseudosigma) of less than 25 percent.

2. Summary of Method

This method is designed for the determination of 14 human-use pharmaceuticals (table 1) in filtered natural-water samples. The method is applicable to those compounds that are (1) efficiently partitioned from the water onto a chemically modified styrene-divinylbenzene resin-based SPE material, (2) can be quantitatively eluted from the SPE material, and (3) can be efficiently ionized by HPLC/MS with electrospray ionization interface operated in the positive ionization mode.

The pharmaceuticals selected for this method are extracted from previously filtered water samples by using disposable polypropylene syringe cartridges that contain 0.5 g of a chemically modified styrene-divinylbenzene resin-based sorbent. A prefiltered water sample of about 1 L is pumped through the SPE cartridge at a flow rate of 15 mL/min. After extraction, the SPE cartridge is dried with air, and the adsorbed compounds are eluted from the dried cartridge by using two sequential elutions of:

(1) 6 mL methanol followed by

(2) 4 mL of methanol, acidified with trifluoroacetic acid (0.1 percent).

Table 1.   Compound names, uses, Chemical Abstract Service registry numbers, and codes for human-use pharmaceuticals determined using this method.

[Y/N, yes/no; CAS, Chemical Abstract Service]

Compound name	Alternative or common name	Reported as an estimated concentration, indicated by an "E" qualifier (Y/N)	Use	CAS registry number[1]	Parameter/ method codes
1,7-Dimethylxanthine	Paraxanthine (metabolite of caffeine)	N	Precursor is a stimulant	611-59-6	6203000021
Albuterol	Salbutamol, Proventil	N	Bronchodilator	18559-94-9	6202000021
Acetaminophen	Tylenol	N	Analgesic	103-90-2	6200000021
Caffeine	Caffenium; Guaranine; methyltheobromide; Methyltheobromine; No-Doz	N	Stimulant	58-08-2	5030500021
Carbamazepine	Epitol; Tegretol; Teril	N	Antiepileptic	298-46-4	6279300021
Codeine	Actacode; Calcidrine; Methylmorphine; N-methyl norcodine; Robitussin AC; Tussar-2; Tussi-organidin	N	Opiate agonist	76-57-3	6200300021
Cotinine	1-Methyl-5-(3-pyridinyl)-2-pyrrolidinone (nicotine metabolite)	N	Precursor is a naturally occurring alkaloid stimulant	486-56-6	6200500021
Dehydronifedipine	Metabolite of nifedipine (Procardia)	N	Precursor is an antiangial	67035-22-7	6200400021
Diltiazem	Dilzem, Tiazac, Cardizem, Cartia XT	Y	Antihypertensive	42399-41-7	6200800021
Diphenhydramine	Banophen, Benadryl, Diphen Af, Genahist, Sleep-eze	N	Antipruritic	58-73-1	6279600021
Sulfamethoxazole	Bactrim; Fectrim; Gantrim; Septrim; Septrin; Sulfotrim; Trisulfam; Urobak	N	Antibiotic	723-46-6	6202100021
Thiabendazole	Bioguard; bovizole; equizole; lombristop; Mintezol; nemapan; omnizole; thiaben	N	Anthelmintic, fungicide	148-79-8	6280100021
Trimethoprim	Abaprim; Chemotrim; Idotrim; Lidaprim; Methoprim; Monotrim; Primosept; Ratiopharm; Trimpex; Uretrim; Wellcoprim	N	Antibiotic	738-70-5	6202300021
Warfarin	Coumadin; Dethnel; Panwarfin; Rodex; Sofarin; Vampirinip; Zoocoumarin	Y	Anticoagulant, rodenticide	81-81-2	6202400021

[1]This report contains CAS Registry Numbers®, which is a Registered Trademark of the American Chemical Society. CAS recommends the verification of the Chemical Abstract Service Registry Numbers through CAS Client Services℠.

The resulting sample extracts are reduced under nitrogen to about 0.1 mL and then reconstituted to a volume of about 1 mL with the initial HPLC eluent, aqueous ammonium formate/formic acid buffer (10 mM, pH 3.7). The pharmaceuticals are chromatographically separated by HPLC by using a reverse-phase octadecylsilane HPLC column, which is coupled to an electrospray ionization interface and quadrupole mass spectrometer for detection, identification, and quantitation. Both internal and external calibration can be used to quantify the pharmaceuticals determined in this method; however, external calibration requires careful attention to final sample extract volumes. The concentrations of 12 of the 14 pharmaceuticals are reported without qualification; the concentrations of two pharmaceuticals are reported as estimates because long-term LRS sample recoveries fall below acceptance criteria for reporting concentrations without qualification. The method detection limits and interim reporting levels for the compounds determined by this method were calculated from recoveries of the pharmaceuticals from reagent water samples amended at 0.05 µg/L, and ranged between 0.0069 and 0.0142 µg/L, and 0.015 and 0.10 µg/L, respectively.

Because the response of the individual pharmaceuticals of interest, the quality-control surrogate compounds, and quantitation internal standard can be suppressed or enhanced by the sample matrix (that is, by matrix effects discussed in detail further in this report), results from several quality-control sample types are necessary to properly interpret the ambient environmental concentrations of pharmaceuticals in aqueous samples. Results from laboratory quality-control samples, including LRS samples and laboratory reagent-blank (LRB) samples, are required to provide insight into the performance of the method in absence of a sample matrix. Two additional field quality-control sample types are required to identify the effect of sample matrix upon aquatic pharmaceutical concentrations. Specifically, matrix-spike recovery samples and replicate environmental samples, collected from representative sample matrix types within the aquatic system under study, are required as a part of the project quality-control plan.

3. Safety Precautions and Waste Disposal

3.1 Conduct all steps in the method that require the use of organic solvents, such as cartridge cleaning, bottle rinsing, cartridge elution, and extract concentration, in a fume hood. Wear eye protection, gloves, and protective clothing in the laboratory area, and when handling reagents, solvents, or any corrosive materials.

3.2 The liquid waste stream produced during sample preparation is about 95 percent water, with the rest of the volume made up of organic solvents and reagents dissolved in water/solvent mixtures. The solvents used are acetonitrile and methanol, and the organic reagents are ammonium formate, formic acid, and trifluoroacetic acid. Note that trifluoroacetic acid is particularly toxic; thus particular attention to good

laboratory practice is required during trifluoroacetic acid use and disposal. Collect the waste stream in thick-walled carboys and dispose according to local regulations for nonchlorinated waste streams. Dispose solvents used to clean or rinse glassware, equipment, and cartridges in the appropriate waste containers. The solid-waste stream produced during sample analysis is composed of used SPE cartridges and assorted glassware (sample vials and pipette tips). Dispose the solid-waste stream according to local regulations.

3.3 Ensure that the electrospray waste exhaust tube and the vacuum pump exhaust tube of the mass spectrometer are vented out of the ambient laboratory atmosphere through ventilation ducting expressly specified for that purpose.

4. Interferences

A wide range of additional chemical constituents, dissolved organic carbon, and matrix components are likely to be retained on, and subsequently eluted from, the SPE sorbent from the water sample. This results, in turn, in potential interferences to the process of efficiently isolating and accurately identifying the selected pharmaceuticals when using this method.

This method is purposely designed for the determination of an array of pharmaceuticals that comprise a wide range of chemical characteristics and elemental and functional group compositions. This design choice militates against uniformly high recoveries from SPE isolation and elution for all the pharmaceuticals studied. Other commonly used approaches for analyte isolation, which result in the retention of one or two chemically similar classes of compounds and result in the decreased retention of other classes, would not suffice to provide the multiple chemical-class pharmaceutical data this method produces.

During extraction, additional compounds that are not of interest and other matrix components may be retained from the sample and can result in at least four different means of method interference. First, coextracted interferences can be sorbed onto the SPE surface, thereby reducing the efficiency of sorption of the selected pharmaceuticals. Similarly, the presence of these additional matrix constituents attached to the sorbent surfaces can interfere with the ability of eluting solvents to adequately remove selected pharmaceuticals during the elution process.

The second means of matrix interference that may affect analysis using this method can occur during instrumental analysis. The presence of large quantities of coextracted interferences present in the final sample extract may result in poor or irreproducible compound retention and decreased ability to chromatographically resolve closely eluting pharmaceuticals on the analytical column. Deviations from expected chromatographic results have been observed as "shifts" from expected retention times and as deformations of the preferred narrow Gaussian peak shape of the chromatographed compounds.

The third means by which interferences potentially present in the method may affect the analysis follows chromatographic separation, during electrospray ionization of the compound. Complex sample matrix constituents that cannot be chromatographically resolved from the selected pharmaceutical peak may produce one or more of the ions that are characteristically produced by the pharmaceutical, which may change the ion area ratios used for identification. The quadrupole mass spectrometer used for this method operates at unit mass resolution. The ability to discriminate between ions with similar, though not exact, mass-to-charge ratios is limited to, under the best of conditions, 0.1 atomic mass unit, and more typically, 0.5 atomic mass unit of difference between ions of similar mass-to-charge ratios. When the chromatographic peak of the selected pharmaceutical contains coeluting matrix interferences, this can result in ions from matrix interferences to be interpreted as produced by the selected pharmaceutical. The result would be ion-area ratios that deviate substantially from the expected ratio for the compound of interest. This result alerts the analyst to the probability that ions from another interfering compound have been detected and included in the compound mass spectrum because the mass spectrometer was not able to discriminate between them.

The fourth means by which interferences potentially present in the method may affect the analysis is by competing for available charge during electrospray ionization of the compound. This competition for available charge results in either an apparent enhancement or an apparent reduction of compound concentration because of the effects of unknown matrix constituents competing for charge with the internal standard or pharmaceutical when these compounds are ionized in the ion-source region of the analytical instrument (Furlong and others, 2000).

Electrospray ionization is an electrochemical phenomenon in which analytes in solution compete for excess charge. Given the complex heterogeneous mixture of chemicals in an environmental sample extract, it is not surprising that matrix suppression or enhancement may occur (Enke, 1997), particularly in the presence of the complex heterogeneous mixture typically found in environmental extracts. Careful choice of internal standards and surrogates is necessary to minimize matrix enhancement or suppression effects. The surrogates and internal standards have been evaluated for minimal ionization matrix effects on a small set of complex samples. Nevertheless, matrix effects may be unavoidable, particularly for complex samples, such as wastewater influents and effluents, for which this method has not been fully validated.

Detailed attention to the inclusion of quality-control samples in studies using this method is required because of the range of matrix effects that may alter the performance of this method. In addition to the quality-control samples discussed later in this report, the inclusion of one or more matrix-spike samples (where one or more environmental samples are fortified with the method compounds), and processing it along with a corresponding unfortified environmental sample, is required as a means to assess these matrix effects for the water types that are a part of any study. This matrix-spike sample is interpreted along with the other laboratory quality-control samples discussed later in this report.

5.   Apparatus and Instrumentation

5.1   Sample Preparation Apparatus and Instrumentation

5.1.1   Automated Sample Extraction Apparatus

5.1.1.1 Cartridge conditioning vacuum SPE extraction manifold, Supelco, Inc. (cat. no. 57030U), Visiprep™ solid-phase extraction vacuum manifold or equivalent, capable of holding 12 sample cartridges.

5.1.1.2 Elution vacuum SPE extraction manifold, International Sorbent Technology (IST) VacMaster™ or equivalent, fitted with adjustable-flow Luer inlet connectors and internal tube rack for 16-mm tubes.

5.1.1.3 Evaporative concentrator, temperature controlled to 40°C and nitrogen gas pressure of 34.47 kPa (5 lb/in²), Zymark Turbo-Vap or equivalent.

5.1.1.4 SPE workstation, Caliper Life Sciences AutoTrace™ extraction workstation 1.20 or equivalent. Two of these systems are used concurrently to prepare a typical set of 10 environmental samples plus an LRS sample and an LRB sample.

5.1.1.5 Nitrogen-driven Venturi vacuum pump with regulator, PIAB Lab Vac™ H40 (cat. no. H40K6-REAC) or equivalent.

5.1.1.6 Vortexing mixer, Vortex Genie, Scientific Industries or equivalent.

5.1.2   Liquid-Handling Apparatus

5.1.2.1 Pipettor, Rainin EDP-Plus™ 10- to 100-μL variable volume electronic pipette (cat. no. EP-100), and Rainin 100- to 1,000-μL variable volume liquid end (cat. no. 6100-069) or equivalent.

5.1.2.2 Wash bottle, VWR low-density polyethylene "squeeze" bottle (cat. no. 16650-107) or equivalent. This bottle is used to dispense organic-free water (section 6.1.4).

5.1.2.3 Adjustable-volume bottle-top liquid dispenser, BrandTech Scientific Dispensette™ bottle-top dispenser (cat. no. 4701131) or equivalent. One each for dispensing methanol (section 6.1.2) and acidified methanol solution (section 6.2.1).

5.1.3   Auxiliary Apparatus

5.1.3.1 Muffle furnace, capable of two-stage temperature increase that can be properly ventilated (NEY 2-2350 Series II or equivalent). The furnace is used for baking Pasteur pipettes, glass fiber filters, and autosampler vials to remove organic contaminants.

5.1.3.2 Filtering apparatus, Geo Tech (cat. no. 83150007 with clamp 17500004) for use with 14.2-cm filters.

5.1.3.3 Pump, ceramic-piston, valveless, capable of pumping 0 to 25 mL/min, Fluid Metering Inc., model QSY-2 CKC or equivalent.

5.1.3.4 Water Purification System, Solution 2000 water purification system, Aqua Solutions®, Inc., Jasper, Ga., model 2002AL or equivalent.

5.2 Sample Analysis Apparatus and Instrumentation

5.2.1 Instrumentation and Computer Hardware/Software

5.2.1.1 Liquid chromatograph, Agilent Technologies 1100 Series liquid chromatographic system, including a 100-position random-access autosampler equipped with a cooling module, a heated column oven, a binary solvent delivery system, and an Agilent 1100 Series liquid chromatograph/mass selective detector (Agilent LC/MSD) with an electrospray ionization interface capable of operating in positive ionization mode or equivalent.

5.2.1.2 Instrument control/data acquisition software, Agilent Technologies LC/MSD Chemstation™, Revision A.10.01 or higher computerized instrument control software installed on a desktop workstation computer, for data-acquisition/reprocessing system or equivalent.

5.2.1.3 Thru-Put Systems, Inc. Target™ Revision 4.0 chromatographic analysis software, data reprocessing software or equivalent.

5.2.2 Chromatographic columns, MetaChem Technologies, Inc., Metasil Basic octadecylsilane (ODS-3), 5-μm particle size; 2.1-mm inside diameter by 150-mm stainless-steel column or equivalent.

6. Reagents and Consumable Materials

NOTE: Material Safety Data Sheets for all materials described herein should be read prior to using any of these materials to ensure safe handling and proper disposal. Unless otherwise specified (that is, "Standards," section 7), store solutions at room temperature and discard after 6 months. At the NWQL, all solutions are labeled in accordance with the NWQL Quality Management System, section 3.5.1 (Maloney, 2005).

6.1 Neat Reagents

6.1.1 *Liquinox, liquid detergent*—Alconox Inc. or equivalent.

6.1.2 *Methanol*—Burdick & Jackson (cat. no. 230), HPLC grade or equivalent.

6.1.3 *Trifluoroacetic acid (TFA)*—Pierce Chemical, Inc. (cat. no. 28903), Sequanal grade or equivalent.

6.1.4 *Water, organic-free*—Deionized and distilled water that is free from interfering organic compounds and chlorine. Water of appropriate quality is produced by a Solution 2000 water purification system (Model 2002AL, Aqua Solutions, Inc., Jasper, Ga.) or equivalent (section 5.1.3.4).

6.1.5 *Ammonium formate*—96-percent minimum assay JT Baker (cat. no. M530-08) or equivalent.

6.1.6 *Formic acid*—EM Scientific, 98 percent (cat. no. FX0440-7) or equivalent.

6.1.7 *Acetonitrile*—Ultrapure, suitable for HPLC, Burdick & Jackson ultraviolet (UV) grade (cat. no. 015-4) or equivalent.

6.2 Reagent Solutions

6.2.1 *TFA-acidified methanol SPE cartridge elution solution, 0.1 percent*—Add 100 μL of TFA to 110 mL of methanol. Prepare daily.

6.2.2 *Liquinox detergent solution*—Dilute 4 drops of Liquinox (section 6.1.1) with 4 L of organic-free water (section 6.1.4).

6.2.3 *1-M ammonium formate solution*—Dissolve 65.69 g of ammonium formate (section 6.1.5) in 1 L of organic-free water (section 6.1.4).

6.2.4 *1-M formic acid solution*—Dilute 38.8 mL of formic acid (section 6.1.6) with organic-free water (section 6.1.4) to a final volume of 1 L.

6.2.5 *Formate buffer solution, 10 mM*—Dilute 10 mL 1-M ammonium formate solution (section 6.2.3) and 12 mL 1-M formic acid solution (section 6.2.4) with organic-free water (section 6.1.4) to a final volume of 1 L. The pH of this solution should be about 3.7.

6.2.6 HPLC Eluents

6.2.6.1 *Formate buffer eluent, 10 mM*—This eluent is made identically to the formate buffer solution (section 6.2.5) and is used as the aqueous HPLC eluent for positive mode analysis.

6.2.6.2 *Acetonitrile*—Ultrapure, suitable for HPLC, Burdick & Jackson UV (cat. no. 015-4) or equivalent, used unmodified as organic HPLC eluent. (This is the same solvent as listed in section 6.1.7).

6.3 Consumable Materials

6.3.1 Amber glass bottles, 1,000 mL, fitted with Teflon-lined screw caps or equivalent, baked at 450°C ±10°C for a minimum of 4 hours before use.

6.3.2 Autosampler vials, National Scientific Company (cat. no. C4000-2W), 2-mL, graduated amber glass for use with screw-top caps or equivalent.

6.3.3 Vial caps and septa, National Scientific Company (cat. no. C4000-53B), blue screw-top caps that have 11-mm dual Teflon-faced silicone rubber septa or equivalent.

6.3.4 Nitrogen gas, for sample extract concentration, ultrapure.

6.3.5 Sample extract (test) tubes, VWR Durex™ borosilicate glass, 16×100 mm (cat. no. 47729-576) or equivalent. Bake at 450°C for a minimum of 4 hours before use.

6.3.6 Sample extract (test) tube caps, VWR Safe-T-Flex™ caps, 16-mm inside diameter (cat. no. 60828-768) blue caps.

6.3.7   SPE cartridges, Waters® Oasis® HLB, 500 mg, in 6-mL syringe barrel (cat. no. 186000115).

6.3.8   Filter membrane, 14.2-cm diameter, 0.7-μm glass fiber, Pall Corporation (cat. no. 66257) or equivalent. Bake at 450°C for a minimum of 4 hours before use.

6.3.9   Pasteur pipettes, 14.6 cm (5 ¾ in) or 22.9 cm (9 in). Bake at 450°C for a minimum of 4 hours before use.

6.3.10   Rubber pipette bulbs, to fit onto Pasteur pipettes.

6.3.11   Syringes, 5-mL Luer-Lok™ syringes, Beckton Dickinson BD 5 mL, disposable (cat. no. BD306603) or equivalent.

6.3.12   Syringe filters, Acrodisc CR 13-mm, 0.2-μm polytetrafluoroethylene (PTFE) membrane, HPLC certified syringe filters (cat. no. 4423) or equivalent.

6.3.13   Solid-phase adaptors, empty 6-mL syringe barrels, for SPE workstation (section 5.1.1.4) cleaning. Can be adapted from SPE cartridges (section 6.3.7) or equivalent.

7.   Standards

7.1   *Stock single-component standard solutions at 10,000 ng/μL*—Obtain method compounds and surrogate compounds as neat materials at greater than 99-percent purity, if possible, from commercial vendors. If 99-percent purity is not available, lower purity standards can be used but purity must be known. Individual single-component standard solutions at about 10,000 ng/μL (10 mg/mL) are prepared in methanol (section 6.1.2) from neat material by accurately weighing, to the nearest 0.002 mg, 200 mg of the neat material in a 20-mL volumetric flask and diluted to volume. The mass of neat standard and final stock solution volume can be adjusted according to solution volume requirements or preferences. After formulation, all solutions are stored in a refrigerator at 4°C ±2°C in amber glass vials with Teflon-faced, silicone rubber-lined screw caps (section 6.3.1).

7.2   *Method compound intermediate standard and surrogate spiking intermediate standard solutions*—Prepare two separate intermediate multicomponent solutions: the method compound intermediate standard solution and the surrogate spiking intermediate standard solution. The method compound intermediate standard solution contains all compounds except for surrogate compounds. The surrogate spiking intermediate standard solution contains only the surrogate compounds. The process for each intermediate preparation is identical. In a 100-mL volumetric flask, add aliquots of each intermediate compound (method compound or surrogate) by calculating the aliquot of each individual stock solution necessary to produce a final concentration of 40,000 μg/L, calculated as follows:

$$V_{ss} = C_f \left[\frac{V_f}{C_{ss}} \right] \qquad (1)$$

where

V_{ss} = the stock solution volume used (in microliters);

C_f = the final solution concentration (for this solution 40,000 μg/L);

V_f = the final solution volume (for this solution 100,000 μL);

and

C_{ss} = the stock solution concentration (for this solution ~1×10⁷ μg/L).

Dilute to volume with methanol. Store all solutions in a refrigerator at 4°C ±2°C in baked amber glass vials or bottles with Teflon-lined screw caps. Use intermediate multicomponent standard and surrogate solutions for no more than 1 year before recertification is required to validate concentrations.

7.3   *Method surrogate spiking solution at 5,000 μg/L*—The method surrogate spiking solution is composed of two deuterated compounds, ethyl nicotinate d_4 and carbamazepine d_{10}. Add 1,250 μL of the surrogate spiking intermediate standard solution to a 10-mL volumetric flask, and dilute to volume with methanol to produce a final concentration of 5,000 μg/L. This solution then is transferred to a baked amber glass bottle with teflon-lined screw caps, and stored in a refrigerator at 4°C ±2°C. Use the method surrogate spiking solution for no more than 1 year before recertification is required to validate concentrations.

7.4   *Method compound spiking solution at 2,500 μg/L*—The method compound spiking solution consists of all pharmaceuticals determined by this method that are not surrogates or internal standards (see table 1). This solution is made by combining 1,250 μL of the method compound intermediate standard solution to a 20-mL volumetric flask, and diluting to volume with methanol to produce a solution at a final concentration of 2,500 μg/L. All solutions are transferred to, and stored in, a baked amber glass bottle with Teflon-lined screw caps. This solution then is transferred to a cleaned and baked amber glass bottle with teflon-lined screw caps, and stored in a refrigerator at 4°C ±2°C. Use the method compound spiking solution for no more than 1 year before recertification is required to validate concentrations.

7.5   *Internal standard solution*—The internal standard solution consists of isotopically labeled nicotinamide d_4. First, prepare a stock solution in methanol at a concentration of 20,000 μg/L. Then prepare the final spiking solution by diluting the stock solution to a final concentration of 2,500 μg/L, according to the formula described in section 7.2. The use of internal standard quantitation is optional for this method.

7.6   *Calibration solutions*—Prepare a series of calibration solutions that encompasses the calibration range of the method. For the data contained in this report, the calibration range is between 0.005 and 1.00 μg/L in aqueous environmental samples. Equal volumes of the method compound intermediate standard and the surrogate spiking intermediate standard solutions are combined using volumetric pipettors to produce a 20,000 µg/L calibration stock solution that is then serially diluted in methanol to produce the necessary calibration solutions. Table 2 lists the suggested volumes of calibration stock solution, appropriate volumetric flask volumes, and dilution volumes necessary to produce the calibration standards used in this method. Equivalent aqueous concentrations also are listed in table 2. Equivalent aqueous concentrations are the concentrations expected if 50 µL of each calibration solution were diluted to 1,000 mL (the standard volume of a quality-control or environmental sample), then concentrated to 1 mL (the final volume of a quality-control or environmental sample extract).

7.7 *Solution handling precautions*—Prior to use, all stock intermediate and final solutions for method compounds, surrogate standards, internal standards, and calibration standards need to be brought to room temperature and mixed using a vortex mixer (section 5.1.1.6) to ensure homogeneity. Large volumes of standard solutions that are used over a long period (that is, standard and surrogate solutions used for a year) can be subdivided into smaller (5 mL or less) aliquots. Using small aliquots limits the effects of warming and cooling cycles, which occur during normal use, on solution quality.

8. Sample Preparation

8.1 Sample Filtration

This method is applicable only to filtered water samples. All samples should be filtered in the field, preferably at the time of collection. Filtration will reduce the likelihood of compound degradation by removing particulate-associated bacteria. Removal of particulates also will prevent clogging of the retaining frit and stationary phase of the SPE cartridge,

thus improving operation and extraction efficiency. Wilde and others (2004) describe a USGS-approved filtration procedure appropriate for samples analyzed by this method. Briefly, this procedure uses 14.2-cm diameter, pre-baked glass-fiber filters with a nominal 0.7-µm pore size. A positive displacement pumping system is used to process a 1-L water sample through the filter, which is contained in an aluminum filter holder. All materials are compatible with trace organic sampling and the cleaning procedures used to decontaminate the filtering system between samples.

Occasionally, samples are not filtered on site or become cloudy (particulate formation caused by chemical reactions or nanobacterial growth) during transit to the laboratory. Filter these cloudy samples at the laboratory according to the procedure outlined by Wilde and others (2004) by using a 14.2-cm filter holder and positive pressure pump. Use a 0.7-µm pore size, 14.2-cm diameter, glass-fiber filter, baked at 440°C for at least 2 hours (section 6.3.8). Prior to sample filtration, sequentially flush the filtration apparatus with 100 mL of Liquinox solution (section 6.2.2), 100 mL of organic-free water (section 6.1.4), 50 mL of methanol (section 6.1.2), and again with 100 mL of organic-free water. Repeat this cleaning procedure between samples. Use a separate filter for each sample to prevent sample cross-contamination.

8.2 Solid-Phase Extraction Cartridge Cleaning and Conditioning

NOTE: The extraction and elution procedure used in this method was designed to perform equally well by manual operation or by automated SPE workstations. The AutoTrace automated SPE workstation (section 5.1.1.4) was used in the development of this method. The same cleaning and conditioning procedure is used for both. Procedures for automating this step are specific to each workstation. Consult workstation documentation for equipment-specific instructions on how to automate this SPE procedure.

Table 2. Calibration standard solution concentrations, equivalent aqueous concentrations, and calibration stock dilutions for the calibration standards used in this method.

[µg/L, micrograms per liter; mL, milliliter; N/A, not applicable; µL, microliter]

Calibration standard concentration (µg/L)	Calibration standard equivalent aqueous concentration* (µg/L)	Volumetric flask size used to make calibration standard (mL)	Volume of 20,000-µg/L calibration stock solution necessary to produce final calibration standard concentration (mL)
100	0.005	200	1
200	0.01	100	1
400	0.02	50	1
800	0.04	25	1
2,000	0.10	10	1
4,000	0.20	10	2
8,000	0.40	10	4
20,000	1.0	NA	Use undiluted

*Equivalent aqueous concentrations are the concentrations expected if 50 µL of each calibration standard were diluted to 1,000 mL (the standard volume of a quality-control or environmental sample), then concentrated to 1 mL (the final volume of an environmental or quality-control sample extract) .

8.2.1   Cartridges are conditioned by sequentially eluting the cartridge with two 5-mL aliquots of methanol under gravity flow. This step is followed by elution with 5 mL of organic-free water under gravity flow. In this method, gravity flow conditioning through the SPE manifold (section 5.1.1.1) is used in conjunction with the automated SPE workstation because the automated SPE workstation conditions cartridges serially, whereas the manifold can be used to condition multiple cartridges in parallel, saving time.

8.2.2   Conditioned cartridges are used for sample extraction within a short enough time span to ensure that the sorbent bed does not become dry. In no case should the surface of the sorbent bed be allowed to dry out and become exposed to air. If the sorbent bed dries out, the cartridge is conditioned again according to the procedure in section 8.2.1.

8.3   Solid-Phase Extraction

NOTE: The following description is for the automated SPE workstation method. As noted in section 8.2, this method can be carried out manually through the elution step by using the same conditions outlined in the following procedure.

8.3.1   Obtain up to 10 filtered (section 8.1) environmental water samples for analysis. Prepare laboratory reagent blank (LRB) and laboratory reagent spike (LRS) samples, as follows. Obtain two cleaned and baked 1-L amber bottles. Fill them with 1,000 mL organic-free water. In preparing the fortified LRS sample, 100 µL of method compound spiking solution (section 7.4) is added to one bottle containing 1,000 mL of reagent water. This will result in a final concentration of 0.25 µg/L per analyte in the LRS. Record the solution code and bottle preparation date of the method surrogate spiking and the method compound spiking solutions (sections 7.3 and 7.4). Record the combined sample and bottle weight, in grams, for the environmental, LRB, and LRS samples. Add 100 µL of method surrogate spiking solution to each environmental sample bottle and the LRS and the LRB bottles (section 7.3); this produces a final surrogate concentration of 0.5 µg/L in the LRS and LRB. Shake the bottles to thoroughly mix the components of the added solutions in the water. The LRS and LRB samples are prepared for each set of environmental samples. A set of samples in this procedure consists of the LRS and LRB samples and up to 10 environmental samples. The environmental sample total may include duplicate field samples or field samples that are to be fortified in the laboratory (laboratory matrix-spike samples).

8.3.2   As discussed later in this report under "Results and Discussion of Method Validation," environmental samples need to be extracted no later than 5 days after sample collection to minimize sample changes after collection. Samples are shipped on ice by overnight express, and, until extraction, are refrigerated at about 4°C. Remove environmental samples from refrigeration just prior to extraction and allow to warm to room temperature.

8.3.3   Install six conditioned, 6-mL SPE cartridges (section 6.3.7) on the AutoTrace SPE workstation, ensuring that the sample intake tubes are thoroughly submerged in the sample so that the sample will be completely pumped onto the cartridge.

8.3.4   Pump water samples through the conditioned cartridges using a flow rate of 15 mL/min. Approximate extraction time for 1 L of sample is 70 minutes. The AutoTrace workstation will emit an audible signal and suspend operation when extraction is complete.

8.3.5   Upon completion of extraction, the SPE cartridges retain a small volume of residual interstitial sample water within the cartridge sorbent bed. Prior to elution, that water volume is minimized so as not to interfere with subsequent analyte elution and volume reduction steps by drawing room air through the cartridges by vacuum. This step expels as much retained water as possible from the sorbent bed and evaporates a portion of water adhering to the surfaces of the sorbent bed. To do this, remove the cartridges from the AutoTrace SPE workstation (section 5.1.1.4), and return them to the elution manifold (section 5.1.1.2) used to condition the cartridges prior to sample extraction. Ensure that the vacuum line, capable of creating a vacuum of 400 to 450 mbar, is attached to the manifold. In this study a nitrogen-driven Venturi vacuum pump (section 5.1.1.5) was used as a vacuum source. Adjust the vacuum pump to create a vacuum of 400 to 450 mbar. After 10 minutes, carefully release the vacuum so that the manifold stops drawing room air through the cartridges. The cartridges may retain a small amount (about 0.1 mL) of residual water; this amount of water will not affect subsequent elution and extract-volume reduction steps.

8.3.6   The fluid-flow paths of the AutoTrace workstation consist of polytetrafluoroethylene tubing, which may adsorb nonpolar analytes. Therefore, a post-extraction cleaning of the AutoTrace workstation (or manual SPE extraction apparatus) is required and needs to be performed immediately after elution. Properly discard the SPE sample cartridges after elution is complete, and seal the empty adaptor cartridges (section 6.3.13) into the elution stations, to ensure that wash solutions can be pumped through the AutoTrace lines for each sample. Wash the fluid lines of each AutoTrace concentration and elution station by sequentially pumping 50 mL of Liquinox detergent solution (section 6.2.2), 300 mL of organic-free water (section 6.1.4), and 100 mL of methanol (section 6.1.2) through them at a flow rate of 20 mL/min.

8.3.7   Weigh the empty sample bottle and record the weight. The difference between this weight and the combined sample and bottle weight (section 8.3.1) provides the sample mass in grams, which is assumed to be equal to the sample volume in milliliters. Note that this procedure assumes that the volumetric density of a typical freshwater sample is 1 g/mL. If this method is applied to samples collected from saline environments, determine salinity or density and apply a volume correction.

Occasionally a cartridge will clog, even if a sample has been filtered. This is likely the result of adsorption of coextracted natural organic matter onto the cartridge bed. In this event, the entire sample mass may not have been extracted. Weigh the bottle and remaining sample, discard the remaining sample, and re-weigh the empty bottle. Record these results and this condition. This information is required to accurately determine sample concentration and surrogate recovery. Analyte reporting levels may need to be adjusted in proportion to the amount of sample extracted. A processed sample volume of 700 mL is the minimum required for the results to be reported without adjusting reporting levels. Reporting levels for smaller processed volumes are adjusted proportionately.

8.3.8 SPE Cartridge Elution

8.3.8.1 Borosilicate glass sample extract (test) tubes (section 6.3.5) are labeled with appropriate sample information pertaining to the sample extract they are to contain.

8.3.8.2 The tubes are weighed to the nearest milligram (0.001 g), and the weight is recorded on the sample preparation sheet, laboratory notebook, or other sample analysis documentation.

8.3.8.3 Place the SPE cartridges onto the inlet fittings of the elution manifold (section 5.1.1.2), with the corresponding labeled sample extract tubes (section 6.3.5) positioned in the manifold tube rack directly below.

8.3.8.4 Add methanol (section 6.1.2) in two aliquots of 3 mL each to the barrel of each SPE cartridge being eluted.

8.3.8.5 Using a pipette bulb, apply positive pressure to the top of each SPE cartridge as needed to start the flow of methanol through the sorbent bed.

8.3.8.6 Allow the methanol to flow through under gravity feed into the tubes below.

8.3.8.7 After the full volume of methanol has passed through the sorbent bed, add a 4-mL aliquot of acidified methanol (section 6.2.1) to each cartridge barrel, allowing the entire volume to flow through the sorbent bed under gravity flow.

8.3.8.8 After the flow of acidified methanol ceases, apply vacuum to the outlet connector of the elution manifold (section 5.1.1.2) for several minutes to draw the remaining solvent from the sorbent bed into the tube below. Once solvent droplets are no longer visibly eluting from the cartridge tip, stop applying the vacuum.

8.3.8.9 Position the elution manifold inlet fitting block (lid) over a large waste beaker and clean each inlet fitting used by passing about 0.5 mL of methanol from a squeeze bottle through the inlet fitting and into the waste beaker.

8.3.9 Extract Concentration

8.3.9.1 The sample solvent extracts are reduced in volume to about 100 μL under a nitrogen gas vortex stream. To reduce volume, place the labeled tubes containing the sample extracts into the TurboVap sample concentration apparatus (section 5.1.1.3). Samples are concentrated under a nitrogen gas stream of 35 kPa (5 lb/in^2) while kept at 40°C in a water bath. The volume of the sample extracts is about 12 mL, which nearly fills the 15-mL volume of the sample extract tube. The nitrogen gas pressure, therefore, is increased very slowly from 0 to 35 kPa, while monitoring the behavior of the solvent extracts within the tubes in the TurboVap. This process prevents ejecting droplets or a portion of the extract from the tube into the TurboVap water bath or into other samples. Ensure the tubes are spaced apart as much as possible, at least 3.6 cm, to minimize potential cross-contamination. Note that the settings for bath temperature and nitrogen pressure are optimized for analyte recovery; exceeding the specified temperature or pressure settings will adversely affect compound recoveries. Do not allow extract volume to decrease to less than 100 μL, or analyte recoveries might be adversely affected. Volume reduction typically requires about 90 minutes. When sample volume reduction is achieved, the sample tubes are removed in set order from the TurboVap unit and placed in a rack.

8.3.9.2 If internal standard calibration is used (section 10.2.2), transfer 50 μL of the internal standard solution (section 7.5) to each extract tube by using a calibrated pipettor.

8.3.9.3 Dilute each sample extract to a final volume of about 1 mL with 900 μL of formate buffer solution (section 6.2.5). A test tube with 1 mL of buffer can be used to visually verify that the volume is correct.

8.3.9.4 If external standard calibration is used (section 10.2.2), weigh the tubes containing the sample extracts to the nearest milligram (0.001 g), record the combined weight of the sample and the tube, and subtract the weight of the empty tube on the sample preparation sheet (see Appendix A). Note that the difference in mass calculated is used to determine the final volume of the extract. The extract composition is almost entirely aqueous, and a density of 1 g/mL is assumed.

8.3.10 Sample Filtration and Transfer Into Sample Vials

8.3.10.1 Use pre-baked 2-mL amber screw-top autosampler vials, caps, and septa (sections 6.3.2 and 6.3.3).

8.3.10.2 Label each vial with appropriate lab identification and set numbers. This information is important for evaluating individual sample results by comparison to laboratory set and field quality-control samples (LRS and LRB samples, laborooratory matrix-spike and duplicate field samples), and for correctly tracking samples.

8.3.10.3 Transfer sample extracts to the barrel of a syringe fitted with a 0.2-μm Teflon filter cartridge (section 6.3.12). Filter the extracts by hand pressure into the labeled vial.

8.3.10.4 Cap the sample vials, place in a vial tray, organized by set, and store in a freezer at –20°C until analyzed.

9. Instrumental Analysis

9.1 Instrumental Analysis Overview

The pharmaceuticals contained in sample extracts are separated by using an HPLC fitted with a reverse-phase octadecylsilane column and a water/acetonitrile gradient elution. The fundamentals of HPLC separation are reviewed in Snyder and others (1997), and the reader is referred to this citation if more detailed knowledge of HPLC separation is required. Subsequent to the development of the HPLC separation described in this method, new column-manufacturing techniques have resulted in smaller-sized (1.8-μm rather than 5.0-μm diameter) particles that are used for manufacturing HPLC reversed-phase columns. A reversed-phase HPLC column manufactured using these smaller particles was tested (Rapid Resolution HT column, 2.1×30 mm; 1.8-μm particle diameter, Agilent Technologies, Palo Alto, Calif.) and found to provide equivalent separation of the pharmaceuticals determined with this method, requiring shorter total analysis times and less solvent. Considerable adjustment of elution gradients and other HPLC conditions, and subsequent iterative testing, are necessary to ensure equivalent separations with these smaller particle-size HPLC columns.

The separated components are transported in a flowing stream to the electrospray ionization interface. In the interface, the solvent, pharmaceuticals, and any coeluting components of the sample matrix are nebulized into small droplets and desolvated. During desolvation, compounds are ionized by charge adduction, ion evaporation, or a combination of these ionizing processes. The interested reader can find a detailed discussion of the process of electrospray ionization in Kebarle and Ho (1997).

Coordinated, automated computerized programming is used to control most aspects of chromatographic separation, ionization, fragmentation, ion focusing, mass analysis, detection, and data handling. A typical separation of a standard mixture of the compounds determined under positive ionization conditions is shown in figure 1. Note that coeluting pharmaceutical peaks are not distinguished because this is a reconstructed ion chromatogram of selected-ion monitoring results. These coeluting peaks would be separated and identified by using Target™ Data Analysis Software or equivalent automated graphic data-handling software.

9.2 Instrumental Procedure of HPLC/MS Operation

9.2.1 Sample vials are placed in the autosampler and kept at, or below, 4°C by a Peltier cooling unit (Hewlett Packard/Agilent Technologies 1100 Series HPLC autosampler) to prevent compound degradation. A 5-μL aliquot of the sample extract is injected into the HPLC to start separation.

9.2.2 The analytical separation for the method is achieved by using a reverse-phase octadecylsilane column (section 5.2.2).

9.2.3 Mobile-phase eluents used for HPLC separation are the 10-mM formate buffer solution (section 6.2.6.1) as the aqueous (mobile phase A) eluent and acetonitrile (section 6.2.6.2) as the organic (mobile phase B) eluent.

The mobile-phase composition programming segments for gradient chromatographic separation of method analytes are listed in table 3.

Initial HPLC conditions follow: Autosampler, 4°C; column oven, 27°C; binary mobile phases (mobile phases A and B). The combined mobile phase flow rate is 0.20 mL/min. Each HPLC analysis requires 65 minutes to complete, including a postanalysis column re-equilibration period of 20 minutes. HPLC separation and mass spectrometric (MS) analysis are synchronized by computer control starting at 0.00 minutes. For each combination of HPLC column and HPLC/MS system, specific elution compositions and times are tested iteratively to achieve optimal separation, so the specific times and mobile-phase compositions listed in table 3 need to be used as a starting point for developing an acceptable separation.

Agilent Chemstation™ instrument control software is used for this method to program operational methods, analytical sequences, and to control aquisition settings for HPLC and mass selective detector (MSD) instruments. Equivalent software is available for other instruments and needs to be used to automate sample analysis, data acquisition, and, where appropriate, postacquisition data processing.

The analytical settings for the positive electrospray analysis are listed in table 4. Note that these conditions are specific to the Agilent LC/MSD (section 5.2.1.1). Users of other HPLC/MS systems will need to determine optima for these settings that are specific to their instrument systems.

The specific positive ions monitored for the pharmaceuticals in this method are listed in table 5 and are categorized by windows of retention time that bracket the elution of the pharmaceutical of interest. The ions and the times for which those ions are scanned were optimized for the Agilent LC/MSD, and the specific retention windows and ions associated with each window need to be optimized for every HPLC/MS system.

9.2.4 *Mass spectrometer autotuning*—Prior to any analysis, the mass spectrometer is brought to temperature and gas pressure equilibrium, then tuned to ensure accurate mass assignment and a detector response greater than a manufacturer-specified minimum. Tuning is the process of adjusting MSD conditions to maximize sensitivity, maintain acceptable resolution, and ensure accurate mass alignments. An automated tuning (autotuning) procedure is used, with commercially available tuning solutions for positive ion analysis. The autotuning process is specific to each instrument; refer to the manufacturer-provided documentation for each instrument to determine the specific procedure used to autotune the mass spectrometer. Table 6 lists the ions that must be present and the desired peak width for acceptable tuning in positive mode when using the Agilent tuning solution and autotune procedure supplied with the Agilent LC/MSD. Note that the autotune mass axis calibration must be within 0.2 atomic mass unit (amu). Also note that the appropriate electrometer

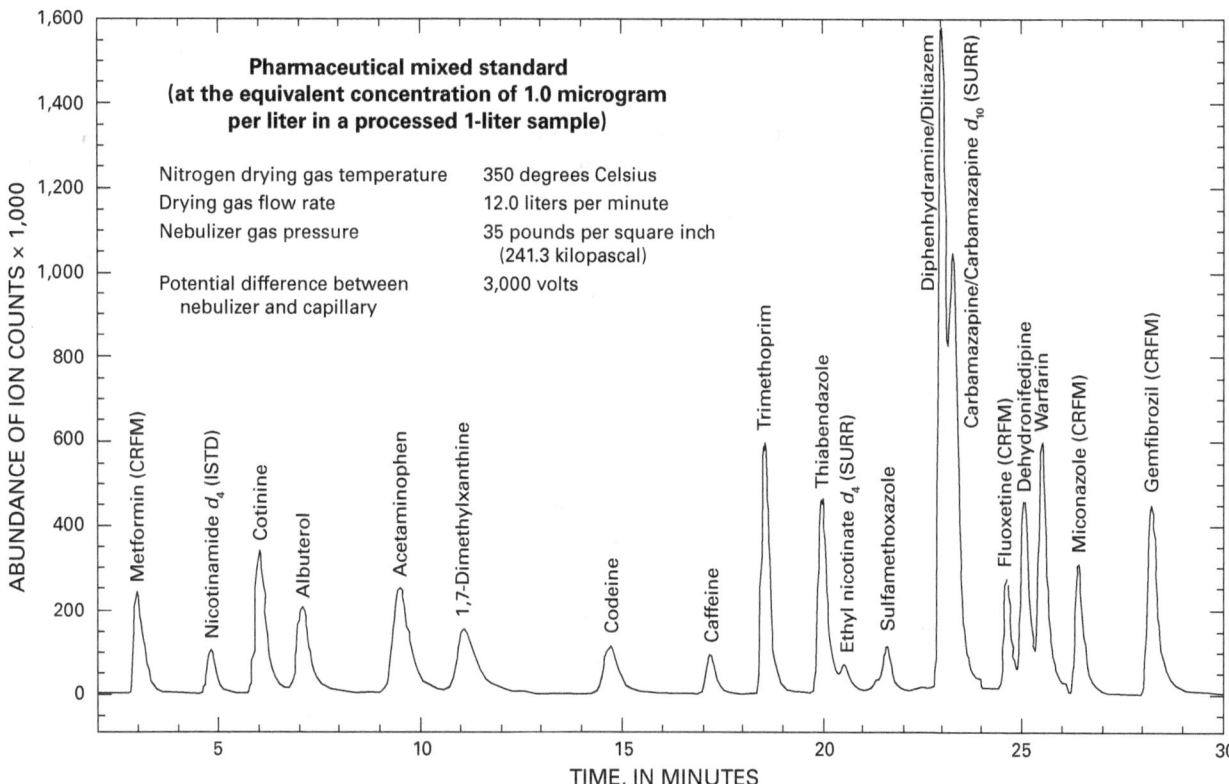

Figure 1. Chromatogram of a positive ion standard used in this pharmaceutical method. ISTD = internal standard compound; CRFM = compound removed from method.

gain (table 5) varies with the condition and age of the mass spectrometer's electron multiplier, and the electrometer gain set during autotune will change with time.

Regardless of the HPLC/MS system used, electronic and paper copies of tuning settings need to be stored in date sequence to monitor long-term HPLC/MS system performance, assist in determining if maintenance or repairs are required, and allow comparison of samples analyzed over extended (greater than 6 months) periods.

The mass spectrometer performance needs to meet manufacturer's specifications for peak width, mass-axis calibration, and minimum acceptable ion signal intensity. If it does, proceed to calibrate quantitatively. However, if the performance is not acceptable, then diagnostic, preventive, or corrective procedures might be required. Manufacturer-supplied diagnostic procedures are used to identify and correct any autotune-identified problems. Following any corrections, repeat the autotune procedure to verify that the corrections result in acceptable instrument performance. When acceptable mass spectrometer peak width, mass-axis calibration, and signal intensity have been achieved, as indicated by a successful autotune, the instrument can be calibrated for quantitative analysis.

9.2.5 *Quantitative calibration*—A multiple concentration calibration for quantitative analysis is carried out for all compounds after acceptable mass spectrometer tuning is completed. The eight concentrations for this calibration are listed in table 7.

Table 3. High-performance liquid chromatograph elution profile for this method.

[A, aqueous mobile phase; B, organic (acetonitrile) mobile phase]

Time, in minutes	Percentage of mobile phase A	Percentage of mobile phase B
0	94	6
5	94	6
9	86	14
10	76	24
15	59	41
16	49	51
26	30	70
27	0	100
39	0	100
45	94	6
65	94	6

Table 4. Electrospray source operating conditions during positive ionization analyses used in this method.

Characteristic	Setting
Nitrogen dry gas temperature:	350 degrees Celsius
Drying gas flow rate:	12.0 liters per minute
Nebulizer gas pressure:	241 kilopascals
Potential difference between nebulizer and capillary:	3,000 volts

Table 5. Mass spectrometer time-programmed operating conditions for individual compounds determined under positive ionization conditions. Note that pharmaceuticals not included in the final method are present in the table.

[*m/z*, mass-to-charge ratio]

Time, in minutes	Group number	Compound names	Selected-ion monitoring ion masses, in *m/z*	Typical electrometer gain	Fragmentor voltage, in volts	Individual ion dwell time, in milliseconds
2.00	Group 1	Metformin	113.0, 130.1	1	125	439
4.20	Group 2	Nicotinamide d_4 (internal standard)	84.0, 127.0	1	125	439
5.55	Group 3	Cotinine	80.1, 98.1,177.1	1	150	145
		Albuterol	166.1, 222.1, 240.1		120	
8.40	Group 4	Acetaminophen	110.1, 152.0	1	120	145
		1-7 Dimethylxanthine	124.1, 181.0		130	
12.80	Group 5	Codeine	215.0, 241.0, 300.1	1	210	292
15.90	Group 6	Caffeine	138.0, 195.1	1	140	145
		Trimethoprim	291.1, 230.1		120	
19.60	Group 7	Ethyl Nicotinate d_4 (surrogate)	128.0, 156.0	1	110	86
		Azithromycin	591.3, 794.4		120	
		Sulfamethoxazole	108.0, 254.0		140	
		Thiabendazole	175.2, 202.0		190	
22.10	Group 8	Diphenhydramine	167.1,256.1	1	100	72
		Carbamazapine	194.1, 237.1		140	
		Diltiazem	178.1, 415.1		150	
		Carbamazapine d_{10}	204.3, 247.3		160	
		Erythromycin	558.3, 716.3		160	
24.00	Group 9	Dehydronifedipine	268.0, 284.0, 345.0	1	190	292
26.15	Group 10	Warfarin	163.0, 309.1	1	100	439
27.35	Group 11	Miconazole	158.9, 416.9, 418.9	1	190	292

The eight calibration concentrations are analyzed sequentially, calibration curves are produced, and from these curves, compound-specific response factors are calculated. A minimum of five quantitation levels, as well as an instrument blank solution, must be analyzed to determine the calibration curve, and the curve should have a correlation coefficient (r^2) greater than 0.995. Most compounds will exhibit linear calibration curves with acceptable r^2 values, but some compounds may require quadratic or exponential curve fitting to achieve acceptable r^2 values. Considerable knowledge and experience are required to determine whether applying quadratic or exponential curve fitting is appropriate. If one or two of the calibration curve standard levels are not used, there has to be a legitimate reason for deleting the point from the curve, such as an incomplete or inaccurate injection, or some evidence that standard solution quality has fallen below acceptable levels. Corrective actions, such as preparing new standards, also may be required.

9.2.6 *Sample analysis*—Prior to quantitative calibration, the HPLC/MS system is verified to ensure it is ready to process calibration and environmental samples. Adequate supplies of the aqueous and organic eluents in the HPLC eluent reservoirs are verified, and in the case of the Chemstation™ instrumental

interface software used in this study, the digital indicators of eluent levels are adjusted, as necessary, to ensure that automated alarms in the software will perform properly.

After the instruments have been equilibrated under initial conditions, the analytical sequence is started from the Chemstation™ instrumental interface. The HPLC/MS will perform the chromatographic separation and acquire analytical data using settings contained in the analytical method specified in the sequence file. The sequence generated includes two injections of instrument blanks to thoroughly rinse and equilibrate the HPLC lines and column prior to analyzing the calibration samples. If the initial calibration is acceptable, the set(s) of environmental

Table 6. Autotune characteristics for acceptable positive ion tuning criteria for the electrospray tuning solution (Agilent part number G2431A) used in this method.

[*m/z*, mass-to-charge ratio; amu, atomic mass unit]

Tune ion mass (*m/z*)	Target peak width (amu)
118.08	0.65
622.03	.65
922.05	.65
1,521.95	.65
2,121.95	.71

Table 7. Working standard concentrations and equivalent aqueous sample concentrations for this method.

[µg/L, microgram per liter; mL, milliliter; NA, not applicable; µL, microliter]

Standard number	Equivalent aqueous concentration* (µg/L)	Working standard solution concentration (µg/L)	Volumetric flask size used to make calibration standard (mL)	Volume of 20,000-µg/L calibration stock solution necessary to produce final working standard concentration (mL)
1	0.005	100	200	1
2	.01	200	100	1
3	.02	400	50	1
4	.04	800	25	1
5	.10	2,000	10	1
6	.20	4,000	10	2
7	.40	8,000	10	4
8	1.0	20,000	NA	Use undiluted

*Equivalent aqueous concentrations are the concentrations expected if 50 µL of each working standard were diluted to 1,000 mL (the standard volume of a quality-control or environmental sample), then concentrated to 1 mL (the final volume of an environmental or quality-control sample extract).

samples, set quality-control (QC) samples, and instrument QC samples that are in the analytical sequence, or batch, are analyzed. A batch typically consists of up to six environmental sample sets of 10 samples each, the associated set QC samples, and instrumental QC samples to monitor performance. Note that analysis of the entire analytical batch can require up to 72 hours of continuous instrument operation. As a result, instrument QC sample data needs to be reviewed during the analysis of the batch to ensure acceptable instrument operation throughout the analysis.

To reduce the potential of losing useful sample data if the operation of either the mass spectrometer or the HPLC fail during a batch, instrumental QC samples are interspersed among environmental and set QC samples in the batch. Review of the results from the interspersed instrumental QC samples while analysis of the batch is ongoing ensures acceptable instrument performance during the analysis of a sample batch and permits partial use of the batch data set if calibration problems or instrument failure occur during analysis.

The first instrument QC sample type used to monitor batch performance is the continuing calibration blank (CCB) sample. CCB samples are used to monitor possible cross-contamination between injections as a result of incomplete injection or insufficient injection-needle washing. A continuing calibration verification (CCV) sample follows each CCB, and ensures ongoing acceptable calibration performance during analysis of the batch. Particular attention is paid to CCB and CCV samples to ensure that contamination between injections is absent and that instrument calibration meets criteria, respectively. An initial calibration curve is calculated after calibration standards in the sequence have been processed. The CCV samples are analyzed after calibration, and the calculated CCV results are used to ensure that the calibration used for environmental samples meets acceptance criteria (Maloney, 2005).

Should either CCB or CCV samples fail to meet acceptance criteria, the analytical sequence is halted and corrective action taken. CCV data are reviewed qualitatively

and quantitatively to ensure that (1) detected compounds are correctly identified, and (2) the instrument responses for the quantitation ions are acceptable. When CCV results exceed statistically derived control limits (Maloney, 2005), a new calibration is required. Typical batch analytical sequences are listed in table 8. Note that this sequence includes analyses for producing the initial calibration curve. If CCV results indicate that calibration meets acceptance criteria, the sequence would be adjusted to allow additional environmental, laboratory QC, and sequence QC samples using the same calibration. Note also that the position of the laboratory reagent blank (LRB) sample in the preparation set is varied to monitor for position-specific contamination. As a result, the position of the LRB sample can vary within the instrument analytical sequence.

10. Calculation of Results

10.1 *Overview of evaluation of analytical sequence results*—Postanalysis data for the analytical sequence are processed using chromatographic data reprocessing software (Target software, section 5.2.1.3), with detailed review performed by the analyst that consists of the following steps: (1) verify acceptable qualitative identification and instrument response during the analytical sequence; (2) produce a calibration table that can be applied to all qualitatively identified pharmaceutical detections in environmental samples; (3) verify any qualitatively identified detections, and ensure all qualitatively identifiable detections were made by the automated software; (4) verify that the integrated quantitation ion response areas (peak areas) for any environmental sample detections are correct; (5) verify the concentrations produced from application of the calibration table to the environmental results are correct; and (6) evaluate the qualitative and quantitative results for laboratory QC samples processed concurrently with environmental samples. These steps are all necessary to ensure results of acceptable quality.

Table 8. Typical minimum sample sequence for high-performance liquid chromatography/mass spectrometry analysis in this method.

[µg/L, microgram per liter; #, number]

Injection number	Quality-control or environmental-sample type
1	Continuing calibration blank (in this method, aqueous buffer solution, or alternatively, pure water)
2	Continuing calibration blank
3	0.005 µg/L concentration standard
4	0.01 µg/L concentration standard
5	0.02 µg/L concentration standard
6	0.04 µg/L concentration standard
7	0.10 µg/L concentration standard
8	0.20 µg/L concentration standard
9	0.40 µg/L concentration standard
10	1.00 µg/L concentration standard
11	Third-party check standard (for this method a 0.175-µg/L concentration is used)
12	Continuing calibration verification standard (for this method, a 0.20-µg/L concentration is used)
13	Continuing calibration blank (ultrapure solvent)
14	Environmental sample—#1
15	Environmental sample—#2
16	Environmental sample—#3
17	Environmental sample—#4
18	Environmental sample—#5
19	Environmental sample—#6
20	Environmental sample—#7
21	Environmental sample—#8
22	Environmental sample—#9
23	Environmental sample—#10
24	Set quality-control sample—#11 (typically laboratory reagent blank, but can vary in sequence position)
25	Set quality-control sample—#12 (typically laboratory reagent spike at fortified concentration of 0.25 µg/L)
26	Continuing calibration verification standard
27	Continuing calibration blank
28–39	Twelve sequence entries for environmental/quality-control samples (#13–#24)
40	Continuing calibration verification standard
41	Continuing calibration blank
42–53	Twelve sequence entries for environmental/quality-control samples (#25–#36)
54	Continuing calibration verification standard
55	Continuing calibration blank
56–67	Twelve sequence entries for environmental/quality-control samples (#37–#48)
68	Continuing calibration verification standard
69	Continuing calibration blank
70–81	Twelve sequence entries for environmental/quality-control samples (#49–#60)
82	Continuing calibration verification standard
83	Continuing calibration blank
84–95	Twelve sequence entries for environmental/quality-control samples (#61–#72)
96	Continuing calibration verification standard
97	Limit of quantitation standard (for this method, 0.02 µg/L is used)
98	Continuing calibration blank

During sample analysis, the CCV and CCB samples interspersed in the sequence are evaluated to verify that instrument calibration met criteria throughout the sequence, and that contamination between injections did not occur (section 9.2.6). Because the CCV and CCB samples would have been initially evaluated during the sequence, failure for CCV and CCB to meet acceptance criteria should be infrequent. The processing software calibration table file contains the compound mass–quantitation ion response data for each pharmaceutical determined in the method, the method surrogate compounds, and the internal standard. The data reprocessing software uses this information, as well as the concentrations of the standards, the sample volume, and the compound-specific calibration curves for each pharmaceutical (section 9.2.5), to calculate a concentration for each detected pharmaceutical or surrogate compound in each sample.

If desired, the calibration data produced using this method can be used to determine pharmaceuticals by external standard calibration. External standard calibration may be preferred when sample matrix effects suppress or enhance the response of the injection internal standard. A careful comparison is necessary to determine how precision and accuracy are affected by using either internal or external standard calibrations. When external calibration is used, the calculated final volume of the samples (section 8.3.9.4) must be taken into account in the calculation of analyte concentrations in the samples.

10.2   *Identification and quantitation of pharmaceuticals in environmental samples*—Environmental and QC sample concentrations of pharmaceuticals are determined after an acceptable calibration curve is produced for each pharmaceutical. The determination of a pharmaceutical in an environmental or QC sample is a two-step process in which the compound is qualitatively identified and once identity is established, a quantitative concentration is determined.

10.2.1   *Qualitative determination*—Several criteria are used to determine that a compound identification is qualitatively correct. Correct identification is based on (1) the retention time at which the apex of the chromatographic peak elutes from the HPLC into the MS, (2) the presence of up to three compound-specific ions in the selected-ion monitoring mass spectrum (table 9), and (3) the relative abundances of these ions in the selected-ion monitoring signal and mass spectra, as reflected in the ratios of the primary and secondary confirmation ions to the quantitation ion. These criteria are determined from analysis of authentic standards, and they are verified by analysis of standards in each batch, to compensate for possible long-term changes in HPLC/MS systems. Compounds are qualitatively detected when the following criteria are met:

10.2.1.1   *Quantitation ion*—The quantitation ion is a clearly discernable, selected-ion chromatographic peak that is a predominant ion in the mass spectrum and characteristic of

the pharmaceutical being identified. Typically the quantitation ion also is present with minimal interference. The ratio of the quantitation ion integrated peak area (referred to here as "abundance") to the abundances of up to two additional characteristic confirmation ions provides confirmatory ion ratios that are used as qualitative detection criteria.

10.2.1.2   *Confirmation ion(s)*—One or two detectable, characteristic ions produced from the same pharmaceutical. Confirming ions must have integrated peak response maxima that are coincident with (coelute with) the quantitation ion maximum. Small allowances may be made for ion peaks of less than ideal peak shape, or that are of a discontinuous nature caused by instrument electronic "noise." The integrated abundances (expressed as percentage of the area of the corresponding quantitation ion, also referred to as an ion ratio) must match the ion ratio from a standard analyzed under the same instrumental conditions. Expected ion ratios are listed in table 9. To be an acceptable qualitative identification, the quantitation and the confirmation ion or ions must be present, and the ratio of confirmation to quantitation ions of the pharmaceutical in the sample should be within ±20 percent of the absolute ratios obtained for the same compound from the analysis of a standard solution under the same measurement conditions of this method.

10.2.1.3   *Retention time*—The intensities of the characteristic ions of a compound are at a maximum that should coincide within ±0.1 minute of the selected compound's retention time. Expected analyte retention times are listed in table 9. In addition, the maxima of the quantification ion and the primary and

Table 9.   Quantitation and confirmation ions used for the compounds determined in this method.

[Note that the absence of a secondary confirmation ion indicates that only two stable and sufficiently abundant ions were available for quantitation and confirmation; *m/z*, mass-to-charge ratio]

Compound	Retention time (minutes)	Quantitation ion (*m/z*)	Primary confirmation ion (*m/z*)/peak area ratio (percent)	Secondary confirmation ion (*m/z*)/peak area ratio (percent)
1,7-Dimethylxanthine	11.51	181.0	124.1 / 19	
Acetaminophen	9.73	152.0	110.0 / 26	
Albuterol	7.15	240.1	222.1 / 17	
Caffeine	17.22	195.1	138.0 / 16	
Carbamazepine	23.31	237.1	194.1 / 40	
Codeine	14.82	300.1	241.0 / 15	215.0 / 7
Cotinine	6.06	177.1	80.1 / 24	98.0 / 6
Dehydronifedipine	25.50	345.0	284.0 / 22	268.0 / 10
Diltiazem	23.03	415.1	178.1 / 15	
Diphenhydramine	22.96	256.1	167.1 / 59	
Sulfamethoxazole	21.63	254.0	156.0 / 20	
Thiabendazole	19.98	202.0	175.2 / 32	
Trimethoprim	18.57	291.1	230.1 / 5	
Warfarin	26.41	309.1	163.0 / 39	251.1 / 24
Internal standard				
Nicotinamide-d_4	4.98	127.1	84.0 / 5	
Surrogates				
Carbamazapine-d_{10}	23.19	247.0	204.0 / 68	
Ethyl nicotinate-d_4	21.48	128.1	156.1 / 98	

secondary confirmation ions should be within 0.05 minute of each other. Matrix effects and sample-to-sample pH variations can have a substantial influence on liquid chromatographic retention times, thereby resulting in large variations of absolute retention time reproducibility, which also can be compound-dependent. Thus an absolute retention-time criterion is evaluated in comparison to previously analyzed standards and samples, and with cognizance of known problems that can result in retention-time variations.

10.2.1.4   *Spectra*—The identity of each compound is verified by comparing the selected-ion monitoring spectrum of the suspected compound with a reference selected-ion monitoring spectrum obtained from a standard for that compound analyzed in the same batch; and it must meet acceptance criteria for quantitation (section 10.2.1.1) and confirmation (section 10.2.1.2) ions. Considerable operator judgment is required to determine whether the abundances in the selected-ion profiles are appropriate and if the profiles have relative intensities that are consistent with the reference mass spectrum, or if there are contributions to the relative abundances resulting from interferences. Experience and training are necessary to recognize the salient features of individual mass spectra as well as potential interferences. Exercise careful judgment in making a qualitative identification, given the variability inherent in identifying compounds at concentrations less than a microgram per liter in environmental samples. Specific problems that can make environmental pharmaceutical identification difficult include sample components that are not resolved chromatographically and produce ion signals containing more than one analyte. When chromatographic peak shape is not the expected near-Gaussian form (a broadened peak, shoulders, or a valley between two or more maxima), this result strongly suggests that more than one sample component is present, and appropriate evaluation of analyte spectra and correction for sample background ions (background subtraction) should be considered. When analytes coelute, identification criteria can be met, but each analyte spectrum will contain extraneous ions contributed by the coeluting analyte. Using the appropriate functions of the data-processing software to produce baseline-subtracted mass spectra may help in graphically separating the coeluting ions and correctly identifying selected pharmaceuticals.

10.2.2   *Quantitative determinations*—When a compound is qualitatively determined to be present, a quantitative determination of the compound concentration can then be made. Both internal and external standard calibration can be used to quantify the concentrations of pharmaceuticals determined using this method. The steps necessary to use this method with internal standard calibration are described in this report because, unlike external standard calibration, internal standard calibration requires mixing an internal standard solution and adding an aliquot of that solution to the final sample extract prior to instrumental analysis.

The use of external standard calibration requires strict attention to, and control of, sample extract final volume and injection volume, because the accuracy of these volumes will strongly influence the accuracy of the final concentration.

In contrast, the use of an internal standard compensates for variations in these steps. However, the potential for matrix suppression or enhancement of the internal standard ions during electrospray ionization may be a factor in choosing external calibration. Differences in the magnitude of suppression between the internal standard ions and the ions of the pharmaceutical being quantified can introduce greater error than if external standard calibration were used. The user of this method has the opportunity to evaluate both the internal and external standard procedure for their samples because external standard calibration can be used even if the internal standard has been added. The most reliable means to determine whether internal or external standard calibration should be used routinely is through the analysis of laboratory matrix-spike samples that are representative of the sample types to which the method is applied. Regardless of the calibration method used, data reports describing the results need to indicate whether internal or external standard calibration is used for quantitation of pharmaceutical concentrations.

The concentration of a qualitatively identified pharmaceutical will be based on the integrated area from the primary quantitation ion of that compound, the regression line fitted to the initial calibration curve, and, in the case of internal standard quantitation, the area of the internal standard in the sample, and the internal standard response factors relative to the internal standard response factor from the calibration standards. In practice, linear, quadratic, and exponential curves can provide equally acceptable results; the choice of curve-fit routine used for quantitation should be based on fundamental knowledge of mass spectrometer response, including the detector's signal-saturation characteristics, past compound-specific responses, and the observed relations between the amount of each compound and the corresponding response observed in the calibration data set.

Contributions to the quantitation ion signal from compounds present in the sample matrix may make accurate measurement of compound concentration difficult or impossible. In such cases, the laboratory reporting level can be raised or that compound reported with the U-DELETED flag (unable to determine because of interference).

10.3   *Calculations*—In this method, the calculation of the final concentration of a pharmaceutical in a filtered water sample requires multiple steps, as follows:

10.3.1   Calculate the relative response factors for each pharmaceutical from the calibration analyses conducted in section section 9.2.5 by using a best-fit linear regression or quadratic fit model. For the preferred linear regression, rearrange the equation of the linear form $y = mx + b$ to $m = (y-b)/x$ as follows:

$$RRF_c = \left[\frac{\left(\dfrac{area_c}{area_{is}} \right) - b}{\left(\dfrac{amt_c}{amt_{is}} \right)} \right] \qquad (2)$$

where

RRF_c = the relative response factor for the phamaceutical of interest;

$area_c$ = the integrated peak area of the pharmaceutical of interest;

$area_{is}$ = the integrated peak area of the internal standard used for the pharmaceutical of interest;

amt_c = the mass of the pharmaceutical of interest, in nanograms;

amt_{is} = the mass, in nanograms, of the method internal standard solution (see section 7.5) used for the pharmaceutical of interest;

and

b = the y-intercept of the best-fit linear regression line.

NOTE: A similar calculation can be made for fitted quadratic curve calibrations by rearranging the equation $y = ax^2 + bx + c$, where a, b, and c are experimental constants determined from the fitted curve by iterative mathematical extraction with curve-fitting software.

10.3.2 Calculate the volume of water extracted, in liters (V_s):

$$V_s = (V_i - V_f) / 1{,}000 \qquad (3)$$

where

V_i = initial weight of sample and sample bottle, in grams ($\equiv$mL; section 8.3.1);

V_f = final weight of sample and sample bottle, in grams ($\equiv$mL; section 8.3.7);

and

1,000 = conversion factor for milliliters to liters.

NOTE: This procedure assumes that the volumetric density of a typical freshwater sample is 1 g/mL. For samples collected from saline environments, a salinity or density determination should be made and a volume correction applied (see section 8.3.7).

10.3.3 Calculate sample pharmaceutical concentrations. If the pharmaceutical of interest has met the qualitative identification criteria listed in section 10.2.1, calculate the compound concentration in the sample as follows:

$$C = \left(\frac{amt_{is} \times A_c}{RRF_c \times A_{is} \times V_s} \right) \qquad (4)$$

where

C = the concentration of the pharmaceutical of interest or method surrogate in the sample, in micrograms per liter;

amt_{is} = the mass of internal standard added to the sample, in micrograms;

A_c = the area of the quantitation ion for the pharmaceutical of interest;

RRF_c = the relative response factor for the pharmaceutical of interest, calculated above in section 10.3.1;

A_{is} = the area of the quantitation ion for the internal standard;

and

V_s = the volume of water extracted, in liters, calculated in section 10.3.2 (equation 3).

10.3.4 Calculate the percentage recovery of the surrogate compounds in each sample by using

$$R_a = \left[\frac{C_s}{(C_a \times V_a)/V_s} \right] \times 100 \qquad (5)$$

where

R_a = recovery of surrogate in sample, in percent;

C_s = concentration of surrogate in sample, in micrograms per liter, calculated by using equation 4;

C_a = concentration of compound in the surrogate solution added to the sample (section 7.3), in micrograms per microliter;

V_a = volume of pharmaceutical surrogate solution added to the sample (section 8.3.1), typically 100 µL;

and

V_s = volume of water extracted, in liters (calculated in section 10.3.2).

10.3.5 Calculate the percentage recovery of compounds in the set LRS sample by using

$$R_b = \left[\frac{C_s}{(C_b \times V_b)/V_s} \right] \times 100 \qquad (6)$$

where

R_b = recovery of fortified compound in the set pharmaceutical fortification sample, in percent;

C_s = concentration of compound in set LRS sample, in micrograms per liter, calculated by using equation 4;

C_b = concentration of compound in method compound spiking solution added to sample (section 7.4), in micrograms per microliter;

V_b = volume of reagent spike fortification solution added to the sample (section 8.3.1), typically 100 µL;

and

V_s = volume of laboratory reagent spike sample, in liters (calculated in section 10.3.2).

11. Reporting of Results

11.1 *Reporting units*—Report compound concentrations for field samples in micrograms per liter to four decimal places, but no more than three significant figures. Report data for compounds reported as qualified estimates to four decimal places, but no more than two significant figures. Report surrogate data for each sample type as percent recovered, and report to one decimal place (tenths of a percent), but no more than three significant figures. Report data for the laboratory reagent spike sample as percent recovered, and report to one decimal place, but no more than three significant figures. Compounds quantified in the LRB sample are reported in micrograms per liter to four decimal places, but no more than three significant figures.

11.2 *Reporting limits and levels*—Method detection limits (MDLs) have been calculated for this method by using the procedure outlined by the U.S. Environmental Protection Agency (2005) and are discussed in this report under the section "Results and Discussion of Method Validation." Note that at this time (2008), the U.S. Environmental Protection Agency is reviewing the approach used to determine MDLs, so that the MDLs produced using the current procedure may not be comparable to MDLs determined using revised procedures. Additional information regarding the current status of USEPA MDL calculations can be obtained at *http://www.epa.gov/waterscience/methods/det/*.

The laboratory reporting level (LRL) for each compound determined using this method is calculated according to Oblinger Childress and others (1999) and is twice the MDL. Because qualitatively identified detections that fall below the MDL can provide useful information (Oblinger Childress and others, 1999), report qualitatively identified compound concentrations (those pharmaceuticals that are identified from relative retention time and MS spectral fit) that are less than the MDL or less than the lowest calibration standard as estimated concentrations. Qualitatively identified compound concentrations less than 0.003 µg/L are censored because of the inability to sufficiently discriminate mass spectra signals from instrument or chemical noise. Compounds that are not detected are reported as less than the LRL.

12. Quality Assurance/Quality Control

Laboratory extraction samples are grouped into sets, each consisting of 10 environmental samples, plus LRS and LRB samples, for a total of 12 samples. Field equipment blanks and laboratory matrix-spike samples, whose frequency is determined by the study designer, provide additional quality assurance/quality control (QA/QC). The frequency of analysis of these QA/QC samples and the aspects of the analytical process they monitor are described herein. In addition, appendix B.1 of Maloney (2005) provides a systematic overview of how the NWQL defines and uses results produced from quality-control samples.

12.1 *Surrogate*—Surrogates are organic compounds that are placed into all filtered water samples prior to extraction through the SPE cartridge. Ideally, surrogate compounds are not present in environmental samples, therefore, isotopically labeled analogues of selected pharmaceuticals are preferred choices. In this method, two surrogates are used—ethyl nicotinate-d_4 and carbamazepine-d_{10}. Surrogates are expected to behave similarly to selected pharmaceuticals for SPE recovery and instrumental analysis, and thus they provide an assessment of method performance in environmental samples.

Examination of surrogate recovery for individual samples provides insight into overall method performance for that particular sample. Long-term control limits, determined by using statistical process control techniques on surrogate recoveries from an extended sequence of LRS and LRB samples, are used to evaluate surrogate recoveries of individual samples. If surrogate recoveries fall below performance criteria, then surrogate recovery in the associated environmental samples, LRS, and LRB samples should be evaluated along with any anomalous observations noted during sample preparation.

12.2 *Laboratory Reagent Spike (LRS)*—A 1-L organic-free water sample is fortified at 0.25 µg/L for all pharmaceuticals determined in this method. An LRS sample is included with each sample set and is carried through the entire extraction, elution, and analytical procedure. The LRS recoveries reflect method performance in the absence of any environmental sample matrix. These results are used to determine if overall set recoveries are acceptable, or if there was a gross change in method performance in the set. Individual LRS recoveries are interpreted in the context of a larger data set of LRS recoveries. At a minimum this data set should consist of 30 or more LRS samples, analyzed over a period of 6 months or more, processed by multiple operators, and using more than one instrument for pharmaceutical identification and quantitation. Statistical process control analysis is applied to these data to develop recovery acceptance criteria.

If the recoveries of a set-specific LRS are not acceptable (that is, within two standard deviations of the long-term mean recovery), other measures of set-specific performance, such as surrogate recoveries in the environmental samples and LRB for that sample set, also should be evaluated to determine if there is a set-specific recovery problem. First, any observations recorded during sample preparation for the samples in the set should be reviewed. If it is apparent that poor recovery in the set LRS resulted from a sample processing error, the analyst needs to determine whether the error also adversely affected the environmental samples associated with that set, and if so, corrective action or data qualification should be applied. If surrogate recoveries and internal standard response in the environmental samples and LRB for that sample set are acceptable, then results for the environmental sample detection should be reported; however, these results also should be qualified as estimated concentrations, because the LRS results cannot be used to confirm that the method performed acceptably during the processing and analysis of that set.

12.3 *Laboratory Reagent Blank (LRB)*—A 1-L organic-free water sample is fortified with method surrogates only. One LRB is included with each sample set and is carried through the entire extraction, elution, and analysis procedure. The LRB is used to monitor for interferences and the possible introduction of method pharmaceuticals during sample preparation. If a pharmaceutical is qualitatively identified in an LRB, the possibility that any detections of that pharmaceutical in environmental samples may result from laboratory contamination, either partially or completely, needs to be evaluated. Typically, if a pharmaceutical is detected in a LRB, the concentration is substantially lower than the laboratory reporting level (LRL). Blank detections less than the laboratory reporting level are possible for this method because mass spectrometric analysis can result in a qualitatively identified detection whose concentration is below the statistically derived LRL. For an explanation of below-LRL reporting conventions, see Oblinger Childress and others (1999). At the NWQL, a censoring level of ten times the concentration of the detected compound in the LRB is used to qualify or censor the environmental concentration depending on whether the detection is greater than, or less than, the labotatory reporting level (Maloney, 2005). The concentrations of method compounds detected in LRB samples are not subtracted from those in environmental samples.

12.4 *Continuing Calibration Verification (CCV)*—For each instrumental analytical sequence, a 50-µL aliquot of the 0.20-µg/L calibration standard that contains all of the selected compounds, including surrogates and the internal standard, is added to 900 µL of the aqueous formate buffer solution (section 6.2.5) and 50 µL of the internal standard solution (section 7.5). The mixed solution is added to an autosampler vial, and the vial placed between every 12 environmental and set QA/QC samples throughout the HPLC/MS analysis. These CCV samples are used to ensure that the calibration of the HPLC/MS system is within statistically determined acceptance limits, following the procedures outlined in Maloney (2005). If CCV control limits are exceeded for more than one compound, environmental samples that follow the last CCV that falls within control limits are reanalyzed after appropriate corrective actions and recalibration. If the sample cannot be reanalyzed, results reported for environmental detections of the compounds in question must be qualified as estimated concentrations. Control limits for the CCV rarely were exceeded during the course of this study.

12.5 *Continuing Calibration Blank (CCB)*—A sample of buffer solution (section 6.2.5), or alternatively, organic-free water that contains only the surrogate and internal standard is placed in an autosampler vial to be analyzed

after a CCV. The CCB follows the CCV, and thus monitors for potential injection-to-injection carryover, as well as instrumental contamination.

12.6 *Limit-of-Quantitation (LOQ) Standard*—The LOQ standard is at a concentration approximating the average LRL for the compounds in the method; for the pharmaceuticals determined in this method, that concentration is 0.05 µg/L. The LOQ indicates whether sufficient instrument sensitivity has been maintained throughout the sequence for the determination of low analyte concentrations. The LOQ is analyzed at the end of a sample analytical sequence. If pharmaceuticals at the concentration in the LOQ cannot be qualitatively determined (incorrect ion retention times, ion peaks not seen in signal, incorrect mass spectra), insufficient instrument sensitivity is likely and corrective action is necessary. Any environmental samples analyzed in this set would then be reanalyzed.

12.7 *Field Equipment Blank (FEB)*—A volume of organic-free water is processed exactly as environmental samples by using all appropriate on-site sampling equipment and techniques (Wilde and others, 2004). This process includes bottles, compositing, splitting, and filtering. The FEB is processed at the start of sampling and then about every 15 to 20 samples. The FEB monitors for contamination or carryover, or both, resulting from field sampling and equipment cleaning techniques that could cause equipment contamination of environmental samples. Adhesion of pharmaceuticals to field equipment or sample bottle surfaces is expected to be minimal. For example, triplicate reagent-water, surface-water, and ground-water samples were fortified with 500 ng of the pharmaceuticals determined in this study and held cold for 4 days. The sample bottles were rinsed with solvent after sample analysis to determine if substantial amounts of pharmaceuticals adhere to bottle surfaces during shipment. Eleven of 14 pharmaceuticals were detected in one or more samples; 1,7-dimethylxanthine, acetaminophen, and caffeine were not detected in any of the concentrated bottle wash samples. However, concentrations in the wash samples were low. Mean and median recoveries from the concentrated wash were 0.95 and 0.62 percent, respectively, both less than 5 ng of the 500 ng added. Diltiazem and diphenhydramine were present at the highest concentrations, but the single highest concentrations were 3.5 percent of the 500 ng added, or 17.5 ng, indicating that pharmaceutical adhesion to bottle surfaces was minimal. Additional information about the use of FEB samples can be found in Wilde and others (2004).

12.8 *Laboratory Matrix Spike (LMS)*—The LMS is a duplicate environmental sample that is fortified at 0.25 µg/L for all compounds determined in this method, and is fortified in the laboratory. The unfortified duplicate is used to determine naturally present concentrations of any compounds measured in the sample. If concentrations of method compounds are determined in the unfortified duplicate sample, they are subtracted from the measured concentrations in the fortified sample. The recoveries of method compounds are determined from the background concentration-subtracted results. The LMS measures the effects of the sample matrix on the recovery of method compounds. Several effects are possible, including matrix-enhanced compound degradation, matrix-introduced coeluting interferences, and matrix enhancement of compound concentration. The frequency of LMS analyses is determined by project data-quality objectives. The sample matrices likely to be tested using this method typically are complex, usually contain substantial fractions of wastewater, and matrix-specific effects strongly influence method performance. The inclusion of project-specific LMS samples is required as a means to validate method performance for any particular study. Additional information about matrix-spike samples can be found in Wilde and others (2004).

12.9 *Internal standard performance criteria*—Internal standard response should be evaluated to determine if extract evaporation, ionization suppression or enhancement of the internal standard (matrix effects), or other factors are influencing quantitation. As a rule of thumb, internal standard responses that are less than 50 percent, or greater than 150 percent of the long-term internal standard response in LRS and LRB samples, suggest that detection of any pharmaceuticals in environmental samples be critically evaluated for reanalysis, if a correctable problem can be identified, or reported as estimated concentrations. The effect of a matrix suppression or enhancement is particularly likely if surrogate recoveries exceed control limits.

12.10 *Statistical derivation of quality-control limits*—Long-term control limits for the relevant QC sample types (including CCVs, LRSs, CCBs, LRBs, among others) are derived from data accumulated over an extended period. At the NWQL, this period typically is a calendar year. The data collected and analyzed, and the control limits thus calculated, are compared to previous limits or to initial limits if insufficient previous data exist. There are multiple publications and statistical software packages that can be used for control-limit calculations and interpretation of data for determining correct control limits. These limits need to be stored in such a way that the results from ongoing set QC samples can be readily compared to them.

12.11 *Secondary data review*—A critical component of overall quality assurance for this method is secondary data review. A separate independent chemist who is qualified to perform this analysis reviews all results and documentation to verify that the original analyst correctly identified and quantified the method pharmaceuticals in light of the QC data available and the sample preparation and analysis documentation. The secondary data reviewer ensures that false positive detections, incorrect spectra, incorrect ratios, typographical errors, or other inadvertent errors do not affect the reported results. The results for every environmental sample are subject to secondary data review.

Results and Discussion of Method Validation

Sample Matrix Description

Reagent-water samples and water samples collected from a residential ground-water well and two surface-water sites were used to assess method performance. The sample matrix, as noted in section 4 "Interferences," can adversely affect the determination of pharmaceuticals by this method. Thus, dissolved organic carbon (DOC) concentrations for the water sources used in this study are discussed in the water-source descriptions that follow. DOC concentration is a simple bulk property measure to describe the concentrations of sample organic matrix likely to be present in each water type, and which might affect recovery of pharmaceuticals.

All environmental water samples were fortified at concentrations substantially greater than anticipated ambient concentrations as detected in previous environmental analyses. Fortification as described above may reduce the uncertainty in compound recovery determinations resulting from ambient environmental contributions of either the pharmaceuticals of interest or matrix interferences that may suppress compound ionization.

The pharmaceutical-free, dissolved organic carbon-free reagent water was produced at the NWQL by using a Solution 2000 water purification system (section 5.1.3.4). Briefly, laboratory-deionized water is introduced into a 1-μm activated carbon prefilter, passed through a series of ion exchange resin beds to further reduce conductance to less than 18 megohms and remove dissolved inorganic constituents, followed by high-intensity UV radiation oxidation to remove DOC, and filtered through a 0.22-μm sterile filter. The organic-free water was dispensed into pre-baked 1-L amber bottles for analysis. This water also was used for extraction set QC samples (LRS and LRB samples). Routine monitoring in the water produced by this system showed the DOC concentration to be consistently less than 0.016 mg/L.

The ground-water sample was collected from a single-family domestic supply well near Evergreen, Colorado. Water was collected from the well after a sustained period of domestic use to minimize contributions of water that had been stored in a lined system pressurization tank. The well penetrates 85 m into a fractured rock aquifer with minimal overlying soil. This well (USGS ID 393459105165701) was part of a joint U.S. Geological Survey–Jefferson County, Colorado, ground-water monitoring program near Evergreen, Colorado (Bossong and others, 2003). Water was collected from a tap into a precleaned, 40-L, stainless-steel container and filtered in the laboratory by using the procedure of Wilde and others (2004). Samples were collected sequentially into individual 1-L bottles for analysis. Multiple DOC measurements of the water from this well were made by Bossong and others (2003). All DOC concentrations of these ground-water samples were less than the 1.5-mg/L reporting level of the method used in that study.

Two surface-water sites were sampled for this assessment to evaluate the influence of wastewater contributions on method performance. Water from one site contained a substantial fraction of treated wastewater effluent from an upstream point discharge and the other site did not. Samples from the two sites were studied because observations made during initial use of this method suggested that wastewater effluent contributions to water samples may result in substantial matrix effects on measured pharmaceutical concentrations.

Surface-water samples were collected from the South Platte River as it passes through metropolitan Denver, Colorado. The water quality of the South Platte River has been extensively studied; Litke and Kimbrough (1998) provide an overview. There are substantial contributions of treated wastewater effluent to the South Platte River from upstream permitted discharges at the site where water samples were collected. Grab samples of South Platte River water were collected in stainless-steel containers of either 10- or 40-L capacity. The containers were washed with soap and water, and sequentially rinsed with water and solvent prior to sample collection. The water samples were filtered by using the procedure of Wilde and others (2004), divided into individual 1-L aliquots, and collected into pre-baked 1-L amber bottles for analysis. Median DOC concentrations for the South Platte River at this site (USGS ID 16704000) from 1993 to 1995 were 5.2 mg/L at Denver, Colorado, measured as part of the National Water-Quality Assessment program and described in Litke and Kimbrough (1998).

Water samples from Boulder Creek, near Boulder, Colorado (USGS ID 06730200), were collected at a point upstream from the Boulder wastewater-treatment plant, but downstream from the City of Boulder and several regulated water diversions and inputs. This site was chosen to represent surface water that does not contain substantial contributions of treated wastewater effluent, although there may be contributions of surface runoff or other water from urban uses. The water chemistry characteristics of Boulder Creek and the effects of wastewater discharge on the watershed have been exensively studied (Murphy and others, 2003; Barber and others, 2006). DOC concentrations in Boulder Creek were about 3.0 mg/L in June and October of 2000 (Barber and others, 2003).

Validation Results

Sets of 10 samples of organic-free water were fortified with method pharmaceuticals at concentrations of 0.05, 0.10, and 0.25 µg/L. The results are listed in tables 10, 11, and 12. No method pharmaceuticals were detected in unfortified reagent-water samples used in the validation study. Mean recoveries of individual pharmaceuticals fortified at 0.05 µg/L ranged between 59.5 percent for diphenhydramine to 123 percent for sulfamethoxazole. The standard deviations of recovery at this fortification level ranged between 4.62 and 9.47 percent. Mean recoveries of individual pharmaceuticals fortified at 0.10 µg/L ranged between 47.4 percent for diphenhydramine and 109 percent for sulfamethoxazole. The standard deviations of recovery for individual pharmaceuticals

at this fortification ranged between 4.74 and 12.5 percent. The lowest overall recoveries were observed for pharmaceuticals fortified at 0.25 µg/L. Mean recoveries of individual pharmaceuticals fortified at 0.25 µg/L ranged between 8.62 percent for warfarin and 110 percent for sulfamethoxazole and diltiazem. The standard deviations of recovery for individual pharmaceuticals at this fortification ranged between 2.33 and 12.2 percent. Dehydronifedipine and warfarin exhibited particularly low recoveries. No changes in the individual components of the procedure (including SPE cartridge lot variations, cartridge capacity, spiking solution, and procedural deviation) explain the overall lower recoveries observed at the 0.25-µg/L fortification, or the specifically low recoveries of warfarin and dehydronifedipine.

Sets of 9 to 12 ground-water samples were fortified with method pharmaceuticals at concentrations of 0.025, 0.10, and 0.25 µg/L. The results are listed in tables 13, 14, and 15. In the water sample used for the 0.025-µg/L fortifications, only caffeine and sufamethoxazole were detected in triplicate unfortified samples, with mean concentrations of 0.0034 and 0.143 µg/L, respectively; for the other two fortifications, method pharmaceuticals were not detected in unfortified ground-water samples. The detections of caffeine and sulfamethoxazole are likely the result of bottle-specific contamination of the unfortified samples in the 0.25-µg/L fortification set during sample extraction or carryover during analysis. Recoveries for this fortification were not corrected for ambient environmental concentrations because the recovery of sulfamethoxazole, at 104 percent, did not reflect the observed ambient concentration. Individual pharmaceutical mean recoveries were more consistent in ground water than in reagent water, ranging between 60.3 percent for diltiazem and 127 percent for caffeine at a fortification of 0.025 µg/L, 58.6 percent for thiabendazole and 111 percent for 1,7-dimethylxanthine at a fortification of 0.10 µg/L, and between 58.6 percent for diltiazem and 124 percent for 1,7-dimethylxanthine at a fortification of 0.25 µg/L. The standard deviations of recovery of individual pharmaceuticals ranged between 2.68 and 10.3 percent at a fortification of 0.025 µg/L, between 2.37 and 14.7 percent at a fortification of 0.10 µg/L, and between 2.29 and 12.5 percent at a fortification of 0.25 µg/L. Inspection of tables 13, 14, and 15 reveals more consistent recoveries in the ground-water sample fortifications, when compared to the reagent-water fortification results in tables 10 through 12. This greater consistency likely reflects the presence of low levels of DOC in the ground-water samples. Furlong and others (2001) observed empirically that the presence of low concentrations of DOC in various water samples resulted in more consistent and greater recovery of polar pesticides extracted by SPE and analyzed by ESI–HPLC/MS. The total amount of DOC retained on the SPE during extraction could contribute to the retention of polar pharmaceuticals on the SPE phase, although this effect would be expected to vary between sample DOC matrices. Additionally, the DOC eluted from the cartridge (and present in the final ground-water extracts) is likely at sufficiently low concentration (compared to surface-water samples) that matrix effects, particularly suppression during ionization, are minimized.

Table 10. Bias and variability data from multiple determinations of the method compounds fortified in organic-free reagent water at 0.05 microgram per liter.

[N, number of determinations]

Compound	N	Mean recovery (percent)	Standard deviation of recovery (percent)	Percent relative standard deviation of recovery (percent)	Median recovery (percent)	Minimum recovery (percent)	Maximum recovery (percent)
1,7-Dimethylxanthine	8	117	6.95	5.94	116	107	128
Acetaminophen	8	118	7.96	6.74	118	102	128
Albuterol	8	113	4.62	4.06	113	107	122
Caffeine	8	115	4.98	4.32	114	107	122
Carbamazepine	8	96.5	5.96	6.17	95.2	90	109
Codeine	8	98.9	7.42	7.51	97.8	87	112
Cotinine	8	92.9	9.47	10.2	94.1	79	110
Dehydronifedipine	8	114	7.34	6.45	113	106	125
Diltiazem	8	68.2	5.95	8.73	68.4	58.8	79.0
Diphenhydramine	8	59.5	7.65	12.8	58.1	51.8	77.0
Sulfamethoxazole	8	123	7.92	6.45	122	110	138
Thiabendazole	8	83.7	8.35	9.98	81.6	74.8	99
Trimethoprim	8	95.2	6.78	7.13	95.1	83.8	106
Warfarin	8	114	6.26	5.50	114.4	104.6	122
Surrogates[1]							
Carbamazepine-d_{10}	8	120	8.23	6.83	118	110	137
Ethyl nicotinate-d_4	8	113	5.63	4.96	112	108	126

[1]Surrogate compounds fortified at 0.5 microgram per liter.

Table 11. Bias and variability data from multiple determinations of the method compounds fortified in organic-free reagent water at 0.10 microgram per liter.

[N, number of determinations]

Compound	N	Mean recovery (percent)	Standard deviation of recovery (percent)	Percent relative standard deviation of recovery (percent)	Median recovery (percent)	Minimum recovery (percent)	Maximum recovery (percent)
1,7-Dimethylxanthine	10	89.6	12.5	14.0	92.2	68.3	103
Acetaminophen	10	61.5	8.11	13.2	63.1	47.7	72.4
Albuterol	10	59.1	9.33	15.8	60.4	43.0	72.0
Caffeine	10	92.8	8.51	9.17	91.6	80.9	113
Carbamazepine	10	84.4	7.43	8.81	83.8	73.9	102
Codeine	10	80.8	6.73	8.33	80.4	69.9	97.4
Cotinine	10	67.5	9.03	13.4	66.8	52.9	80.8
Dehydronifedipine	10	106	10.2	9.63	107	93.9	131
Diltiazem	10	55.1	5.54	10.0	54.0	49.3	69.7
Diphenhydramine	10	47.4	4.74	10.0	45.8	43.1	56.9
Sulfamethoxazole	10	109	12.3	11.3	109	86.7	126
Thiabendazole	10	78.2	7.78	9.95	77.4	70.4	97.7
Trimethoprim	10	83.8	7.77	9.27	81.6	74.8	104
Warfarin	10	104	11.1	10.6	104	86.3	126
Surrogates[1]							
Carbamazepine-d_{10}	10	119	7.38	6.18	122	106	128
Ethyl nicotinate-d_4	10	89.2	6.45	7.22	91.3	76.4	96.0

[1]Surrogate compounds fortified at 0.5 microgram per liter.

Table 12. Bias and variability data from multiple determinations of the method compounds fortified in organic-free reagent water at 0.25 microgram per liter.

[N, number of determinations]

Compound	N	Mean recovery (percent)	Standard deviation of recovery (percent)	Percent relative standard deviation of recovery (percent)	Median recovery (percent)	Minimum recovery (percent)	Maximum recovery (percent)
1,7-Dimethylxanthine	10	31.9	2.33	7.30	31.6	27.6	36.2
Acetaminophen	10	63.2	4.59	7.26	62.6	57.8	73.4
Albuterol	10	69.7	12.2	17.4	68.7	52.3	95.6
Caffeine	10	87.7	6.34	7.23	87.6	78.9	100
Carbamazepine	10	79.1	6.75	8.53	80.2	71.6	92.5
Codeine	10	102	12.0	11.7	102	76.6	117
Cotinine	10	79.8	8.96	11.2	80.4	62.7	90.3
Dehydronifedipine	10	9.66	3.13	32.4	9.58	4.48	15.9
Diltiazem	10	110	7.96	7.26	110	94.4	119
Diphenhydramine	10	97.0	9.75	10.0	97.0	75.9	111
Sulfamethoxazole	10	110	9.84	8.91	110	96.5	130
Thiabendazole	10	38.7	7.82	20.2	38.9	26.2	50.0
Trimethoprim	10	56.1	4.66	8.31	56.2	47.2	63.2
Warfarin	10	8.62	3.28	38.1	8.82	2.36	12.9
Surrogates[1]							
Carbamazepine-d_{10}	10	7.33	2.45	33.5	7.38	2.72	10.2
Ethyl nicotinate-d_4	10	98.1	9.96	10.2	98.7	83.5	118

[1]Surrogate compounds fortified at 0.5 microgram per liter.

Table 13. Bias and variability data from multiple determinations of the method compounds fortified at 0.025 microgram per liter in ground water collected from a residential well.

[N, number of determinations]

Compound	N	Mean recovery (percent)	Standard deviation of recovery (percent)	Percent relative standard deviation of recovery (percent)	Median recovery (percent)	Minimum recovery (percent)	Maximum recovery (percent)
1,7-Dimethylxanthine	9	77.4	4.97	6.42	77.6	68.8	84.0
Acetaminophen	9	99.6	3.78	3.79	98.0	97.2	109
Albuterol	9	71.1	4.94	6.96	69.2	66.0	80.0
Caffeine	9	127	4.49	3.54	127	119	131
Carbamazepine	9	89.2	4.88	5.47	90.4	82.0	96.8
Codeine	9	76.7	4.20	5.48	76.4	72.8	84.8
Cotinine	9	99.1	6.50	6.56	98.0	92.0	110
Dehydronifedipine	9	99.5	2.68	2.70	99.2	96.8	104
Diltiazem	9	60.3	5.48	9.09	60.4	50.0	69.6
Diphenhydramine	8	63.1	3.57	5.66	62.8	56.8	67.2
Sulfamethoxazole	9	104	10.3	9.94	107	86.8	116
Thiabendazole	9	69.0	4.80	6.95	69.6	61.2	76.0
Trimethoprim	9	83.4	4.79	5.74	82.0	78.8	93.6
Warfarin	9	82.5	3.82	4.62	81.6	78.0	89.2
Surrogates[1]							
Carbamazepine-d_{10}	12	91.6	3.37	3.68	92.2	86.1	97.6
Ethyl nicotinate-d_4	12	98.8	5.41	5.47	98.6	90.2	107

[1]Surrogate compounds fortified at 0.5 microgram per liter.

Table 14. Bias and variability data from multiple determinations of the method compounds fortified at 0.10 microgram per liter in ground water collected from a residential well.

[N, number of determinations]

Compound	N	Mean recovery (percent)	Standard deviation of recovery (percent)	Percent relative standard deviation of recovery (percent)	Median recovery (percent)	Minimum recovery (percent)	Maximum recovery (percent)
1,7-Dimethylxanthine	10	111	6.40	5.76	110	99.1	119
Acetaminophen	10	107	9.34	8.72	107	87.2	120
Albuterol	10	92.4	4.77	5.16	94.4	84.0	98.3
Caffeine	10	87.4	3.69	4.23	87.5	80.5	92.6
Carbamazepine	10	67.9	2.55	3.76	68.3	63.7	72.0
Codeine	10	87.5	3.31	3.78	87.6	81.6	92.6
Cotinine	10	68.0	14.7	21.6	66.6	51.6	90.0
Dehydronifedipine	10	104	5.18	4.98	102.6	97.0	113
Diltiazem	10	62.3	4.54	7.29	61.4	56.9	69.1
Diphenhydramine	10	64.5	3.36	5.21	63.4	60.2	69.4
Sulfamethoxazole	10	102	6.29	6.14	102	92.0	112
Thiabendazole	10	58.6	2.37	4.04	57.8	55.9	63.0
Trimethoprim	10	75.4	3.46	4.58	75.2	71.2	81.9
Warfarin	10	109	5.10	4.69	109	100	119
Surrogates[1]							
Carbamazepine-d_{10}	10	102	4.56	4.49	102	91.6	108
Ethyl nicotinate-d_4	10	89.0	5.05	5.67	89.8	78.9	95.7

[1]Surrogate compounds fortified at 0.5 microgram per liter.

Table 15. Bias and variability data from multiple determinations of the method compounds fortified at 0.25 microgram per liter in ground water collected from a residential well.

[N, number of determinations]

Compound	N	Mean recovery (percent)	Standard deviation of recovery (percent)	Percent relative standard deviation of recovery (percent)	Median recovery (percent)	Minimum recovery (percent)	Maximum recovery (percent)
1,7-Dimethylxanthine	10	124	7.21	5.84	122	115	137
Acetaminophen	10	115	6.28	5.45	117	105	122
Albuterol	10	95.6	7.75	8.10	95.4	85.6	106
Caffeine	10	87.7	3.85	4.39	88.3	81.6	92.7
Carbamazepine	10	62.6	2.45	3.92	62.9	58.5	67.4
Codeine	10	89.5	5.14	5.74	89.6	82.8	96.0
Cotinine	10	71.8	12.5	17.3	67.3	60.5	94.4
Dehydronifedipine	10	92.5	5.10	5.51	93.5	85.0	99.3
Diltiazem	10	58.6	3.87	6.62	59.7	51.1	63.5
Diphenhydramine	10	61.5	2.41	3.93	61.5	57.5	65.6
Sulfamethoxazole	10	87.4	3.96	4.53	87.6	80.1	93.3
Thiabendazole	10	59.4	2.65	4.47	60.6	54.9	62.0
Trimethoprim	10	73.4	2.29	3.12	73.1	71.2	79.1
Warfarin	10	91.9	4.86	5.29	90.1	86.2	98.0
Surrogates[1]							
Carbamazepine-d_{10}	10	78.4	2.20	2.80	78.8	74.7	81.3
Ethyl nicotinate-d_4	10	88.0	4.04	4.59	87.8	79.3	94.3

[1]Surrogate compounds fortified at 0.5 microgram per liter.

Because of the observed high frequency of pharmaceutical detections in susceptible surface water (Kolpin and others, 2002), and the importance of wastewater-treatment plant discharges as a source of pharmaceuticals to surface water (Glassmeyer and others, 2005), two surface-water sources were sampled to represent surface water that was minimally-to-heavily affected by wastewater discharge (as a component of total flow). The relative amounts of wastewater present in each of these two samples spans the range of conditions in streams likely to be sampled for pharmaceuticals. Ambient concentrations of method pharmaceuticals present in unfortified samples of the two surface-water types were determined in triplicate, and the results of these determinations are listed in table 16. Cotinine and caffeine were detected in Boulder Creek, but only caffeine could be reliably quantified, at a mean concentration of 0.0331±0.0029 µg/L. In contrast, nine pharmaceuticals were detected in the wastewater-dominated South Platte surface water, ranging in concentrations from 0.0040±0.0006 to 0.109±0.0057 µg/L, for acetominophen and caffeine, respectively.

Sets of 10 samples collected from the South Platte River at Denver were fortified with method pharmaceuticals at 0.10, 0.25, and 0.50 µg/L. The results for these fortifications are listed in tables 17, 18, and 19.

Mean recoveries of individual pharmaceuticals fortified at 0.10 µg/L in South Platte surface-water samples ranged between 14.1 percent for sulfamethoxazole and 167 percent for 1,7-dimethylxanthine. Mean recoveries of individual pharmaceuticals fortified at 0.25 µg/L in South Platte surface-water samples ranged between 24.3 percent for thiabendazole and 146 percent for 1,7-dimethylxanthine. Mean recoveries of individual pharmaceuticals fortified at 0.50 µg/L in South Platte surface-water samples ranged between 16.7 percent for sulfamethoxazole and 141 percent for 1,7-dimethylxanthine. The standard deviations of recovery in South Platte surface-water samples ranged between 1.10 and 29.8 percent at a fortification of 0.10 µg/L, between 2.63 and 13.4 percent at a fortification of 0.25 µg/L, and between 2.62 and 25.2 percent at a fortification of 0.50 µg/L.

Mean recoveries for 1,7-dimethylxanthine were consistently high, at 167, 146, and 141 percent in the 0.10-, 0.25-, and 0.5-µg/L fortifications, respectively. High recoveries greater than 120 percent also were observed for albuterol and acetominophen, but not in all three fortifications (compare tables 17–19). Recoveries for codeine, cotinine, and dehydronifedipine were near or greater than 100 percent, after correction of the codeine and cotinine results for ambient pharmaceutical concentrations in these samples. Conversely, recoveries of sulfamethoxazole from South Platte surface water were consistently low, at 14.1, 30.5, and 16.7 percent in the 0.10-, 0.25-, and 0.5-µg/L fortifications, respectively. Similar low recoveries occurred for thiabendazole (tables 17–19).

The recoveries of 9 of 14 pharmaceuticals were corrected for ambient environmental concentrations (table 16) for the South Platte water samples at all three fortifications. It is unlikely that correction for ambient concentrations results in the low recoveries observed for some pharmaceuticals because there is no consistent pattern within a fortification, or more importantly, between fortifications, where substraction of a constant ambient contribution of a pharmaceutical is expected to have less effect upon the higher concentration fortifications. In addition, pharmaceuticals that did not have detectable ambient concentrations, such as trimethoprim and 1,7-dimethylxanthine, exhibited low and high recoveries, respectively. Matrix competition for the SPE stationary phase, which could either enhance or reduce recovery, is expected to affect recoveries more uniformly, rather than the observed dramatic differences seen between recoveries for specific pharmaceuticals. Thus these high and low recoveries likely result from matrix effects during ionization, such as enhancement or suppression of the quantitation and confirmation ions of the internal standard and pharmaceuticals as they compete with sample matrix for a limited quantity of protons during the ionization process.

Sets of 10 samples were collected at Boulder Creek upstream from the wastewater-treatment discharge and were fortified at concentrations of 0.10 and 0.25 µg/L. Recoveries of these fortifications are listed in tables 20 and 21.

For individual compounds, the mean recoveries in the 0.10-µg/L fortification were higher and more varied than the mean recoveries for the 0.25-µg/L fortification. Mean recoveries of individual pharmaceuticals fortified at 0.10 µg/L in Boulder Creek surface-water samples ranged between 35.8 percent for sulfamethoxazole and 146 percent for 1,7-dimethylxanthine. Mean recoveries of individual pharmaceuticals fortified at 0.25 µg/L in Boulder Creek surface-water samples ranged between 22.1 percent for sulfamethoxazole and 88.4 percent for 1,7-dimethylxanthine. The individual Boulder Creek water samples used for the 0.10- and 0.25-µg/L fortifications were collected from a single large filtered water sample, and processed sequentially, so differences caused by sample collection time and processing cannot explain the observed differences in recovery. Precision for both sets of analyses were similar, as reflected in the standard deviation of recovery in the Boulder Creek surface-water results, which ranged between 0.68 and 6.45 percent at the 0.10 µg/L-fortification, and between 1.44 and 6.76 percent in the 0.25-µg/L fortification. One value was excluded from the calculation of summary statistics for caffeine in the 0.10-µg/L fortification of Boulder Creek surface-water samples (table 20). This result, a recovery of 560 percent, is an outlier, probably caused by contamination during sample collection, processing, or analysis. As a result, this value was not included in the data in table 20, and this contamination during a strictly controlled study emphasizes the need for contamination control when collecting, processing, or analyzing samples where the compound of interest, such as caffeine, is in products that are used daily.

Table 16. Ambient environmental concentrations of method pharmaceuticals measured in triplicate determinations of surface water collected from Boulder Creek and the South Platte River.

[<LRL, less than the laboratory reporting level; Detected, qualitatively identified but concentration so low that result is unreliable; NA, not applicable]

Compound	Boulder Creek		South Platte	
	Mean concentration, in micrograms per liter	Standard deviation of concentration, in micrograms per liter	Mean concentration, in micrograms per liter	Standard deviation of concentration, in micrograms per liter
1,7-Dimethylxanthine	<LRL	NA	<LRL	NA
Acetaminophen	<LRL	NA	0.0040	0.0006
Albuterol	<LRL	NA	<LRL	NA
Caffeine	0.0331	0.0029	0.109	0.0057
Carbamazepine	<LRL	NA	0.0454	0.0036
Codeine	<LRL	NA	0.0054	0.0006
Cotinine	Detected	NA	0.0104	0.0017
Dehydronifedipine	<LRL	NA	<LRL	NA
Diltiazem	<LRL	NA	0.0190	0.0005
Diphenhydramine	<LRL	NA	0.0111	0.0010
Sulfamethoxazole	<LRL	NA	0.0529	0.0072
Thiabendazole	<LRL	NA	<LRL	NA
Trimethoprim	<LRL	NA	0.0077	0.0003
Warfarin	<LRL	NA	<LRL	NA

Table 17. Bias and variability data from multiple determinations of the method compounds fortified in surface water from the South Platte River at 0.10 microgram per liter.

[N, number of determinations]

Compound	N	Mean recovery (percent)	Standard deviation of recovery (percent)	Percent relative standard deviation of recovery (percent)	Median recovery (percent)	Minimum recovery (percent)	Maximum recovery (percent)
1,7-Dimethylxanthine	10	167	18.5	11.1	163	143	214
Acetaminophen	10	120	29.8	24.8	111	99.6	203
Albuterol	10	139	3.64	2.61	140	132	144
Caffeine	10	75.8	27.2	35.9	65.1	54.4	138
Carbamazepine	10	45.6	2.01	4.41	45.4	42.8	48.9
Codeine	10	125	5.72	4.56	126	116	135
Cotinine	10	114	22.9	20.1	126	86.0	136
Dehydronifedipine	10	117	3.29	2.81	118	110	122
Diltiazem	10	48.8	2.60	5.33	48.1	45.7	54.3
Diphenhydramine	10	49.4	1.10	2.22	49.2	47.9	50.8
Sulfamethoxazole	10	14.1	6.95	49.3	13.2	3.17	26.4
Thiabendazole	10	27.5	9.84	35.8	23.3	18.0	45.2
Trimethoprim	10	50.2	2.22	4.43	49.9	46.6	53.9
Warfarin	10	77.7	3.91	5.03	78.1	69.1	82.2
Surrogates[1]							
Carbamazepine-d_{10}	10	54.7	1.64	3.00	54.4	52.4	57.7
Ethyl nicotinate-d_4	10	85.4	2.94	3.44	85.8	81.4	89.7

[1]Surrogate compounds fortified at 0.5 microgram per liter.

Table 18. Bias and variability data from multiple determinations of the method compounds fortified in surface water from the South Platte River at 0.25 microgram per liter.

[N, number of determinations]

Compound	N	Mean recovery (percent)	Standard deviation of recovery (percent)	Percent relative standard deviation of recovery (percent)	Median recovery (percent)	Minimum recovery (percent)	Maximum recovery (percent)
1,7-Dimethylxanthine	10	146	6.84	4.67	147	137	159
Acetaminophen	10	116	6.81	5.89	116	105	124
Albuterol	10	143	5.93	4.14	144	130	151
Caffeine	10	76.4	5.37	7.03	78.4	65.2	83.5
Carbamazepine	10	63.9	3.49	5.46	64.6	58.0	68.6
Codeine	10	124	5.25	4.25	125	114	130
Cotinine	10	109	13.4	12.3	111	84.8	128
Dehydronifedipine	10	130	5.42	4.16	133	119	134
Diltiazem	10	55.0	5.04	9.15	53.6	50.0	67.1
Diphenhydramine	10	54.6	4.01	7.35	54.2	47.8	63.1
Sulfamethoxazole	10	30.5	7.09	23.2	27.7	24.9	48.5
Thiabendazole	10	24.3	9.87	40.7	20.5	17.8	51.2
Trimethoprim	10	60.3	2.63	4.36	60.3	55.3	65.2
Warfarin	10	82.8	4.12	4.97	82.5	75.8	90.2
Surrogates[1]							
Carbamazepine-d_{10}	10	67.2	3.14	4.68	67.6	62.4	72.2
Ethyl nicotinate-d_4	10	89.7	4.30	4.79	88.8	84.4	95.5

[1]Surrogate compounds fortified at 0.5 microgram per liter.

Table 19. Bias and variability data from multiple determinations of the method compounds fortified in surface water from the South Platte River at 0.50 microgram per liter.

[N, number of determinations]

Compound	N	Mean recovery (percent)	Standard deviation of recovery (percent)	Percent relative standard deviation of recovery (percent)	Median recovery (percent)	Minimum recovery (percent)	Maximum recovery (percent)
1,7-Dimethylxanthine	9	141	25.2	17.8	141	86.8	179
Acetaminophen	10	100	14.7	14.7	97.9	77.5	122
Albuterol	10	98.6	15.4	15.7	99.5	71.2	123
Caffeine	9	84.6	11.2	13.2	82.7	67.7	106
Carbamazepine	10	61.6	6.67	10.8	63.5	50.6	70.4
Codeine	10	93.9	12.5	13.3	93.7	67.7	114
Cotinine	10	99.7	11.5	11.5	99.1	78.2	118
Dehydronifedipine	10	107	18.1	16.9	104	76.4	136
Diltiazem	10	50.7	5.51	10.9	52.8	40.6	56.1
Diphenhydramine	10	52.0	6.05	11.6	52.2	41.1	61.9
Sulfamethoxazole	10	16.7	2.62	15.7	16.9	11.4	20.2
Thiabendazole	10	28.7	6.77	23.6	28.4	18.8	40.6
Trimethoprim	10	48.5	5.40	11.1	48.4	38.1	55.7
Warfarin	10	85.9	7.61	8.86	87.2	71.3	94.2
Surrogates[1]							
Carbamazepine-d_{10}	12	61.4	6.25	10.2	62.4	51.8	81.3
Ethyl nicotinate-d_4	12	86.8	8.55	9.85	86.3	73.0	94.3

[1]Surrogate compounds fortified at 0.5 microgram per liter.

Table 20.   Bias and variability data from multiple determinations of the method compounds fortified in surface water from Boulder Creek at 0.10 microgram per liter.

[N, number of determinations]

Compound	N	Mean recovery (percent)	Standard deviation of recovery (percent)	Percent relative standard deviation of recovery (percent)	Median recovery (percent)	Minimum recovery (percent)	Maximum recovery (percent)
1,7-Dimethylxanthine	10	146	4.91	3.36	145	140	157
Acetaminophen	10	80.1	3.03	3.78	79.7	75.5	84.3
Albuterol	10	75.1	2.77	3.69	76.2	69.5	77.9
Caffeine	9	42.9	6.45	15.0	41.5	35.7	52.6
Carbamazepine	10	49.6	1.70	3.42	49.1	47.4	53.2
Codeine	10	62.1	1.55	2.50	62.2	60.0	64.3
Cotinine	10	60.0	5.17	8.61	59.5	51.6	66.9
Dehydronifedipine	10	61.4	3.43	5.60	60.8	56.3	67.8
Diltiazem	10	44.9	1.40	3.11	44.8	43.0	46.7
Diphenhydramine	10	44.2	1.02	2.31	44.3	42.4	45.6
Sulfamethoxazole	10	35.8	1.66	4.64	35.0	34.3	39.3
Thiabendazole	10	36.2	3.48	9.60	34.8	32.6	42.5
Trimethoprim	10	49.7	0.68	1.36	49.5	48.8	50.9
Warfarin	10	56.5	1.96	3.47	56.8	53.0	59.0
Surrogates[1]							
Carbamazepine-d_{10}	10	44.9	1.68	3.75	44.9	41.2	47.2
Ethyl nicotinate-d_4	10	65.5	4.03	6.14	66.0	60.2	71.1

[1]Surrogate compounds fortified at 0.5 microgram per liter.

Table 21.   Bias and variability data from multiple determinations of the method compounds fortified in surface water from Boulder Creek at 0.25 microgram per liter.

[N, number of determinations]

Compound	N	Mean recovery (percent)	Standard deviation of recovery (percent)	Percent relative standard deviation of recovery (percent)	Median recovery (percent)	Minimum recovery (percent)	Maximum recovery (percent)
1,7-Dimethylxanthine	10	88.4	4.12	4.65	87.6	83.4	95.7
Acetaminophen	10	62.5	6.76	10.8	62.1	51.8	73.0
Albuterol	10	60.1	4.72	7.86	60.2	54.0	68.3
Caffeine	10	54.6	5.80	10.6	54.1	44.1	66.0
Carbamazepine	10	39.6	2.46	6.21	39.8	36.2	45.2
Codeine	10	48.5	3.41	7.02	48.3	43.4	56.0
Cotinine	10	47.3	5.07	10.7	48.3	39.4	55.1
Dehydronifedipine	10	78.5	6.30	8.02	78.7	67.8	91.5
Diltiazem	10	32.3	3.51	10.9	31.9	25.6	39.6
Diphenhydramine	10	31.2	3.22	10.3	31.0	24.6	37.4
Sulfamethoxazole	10	22.1	1.44	6.52	21.9	20.5	25.6
Thiabendazole	10	23.1	2.15	9.31	23.1	19.3	26.7
Trimethoprim	10	36.6	2.34	6.40	36.6	33.2	41.6
Warfarin	10	42.2	2.54	6.01	42.1	39.0	48.0
Surrogates[1]							
Carbamazepine-d_{10}	10	42.3	3.60	8.52	42.5	36.2	49.7
Ethyl nicotinate-d_4	10	64.2	5.95	9.26	63.9	54.4	75.5

[1]Surrogate compounds fortified at 0.5 microgram per liter.

Comparison of Validation Results

Surface-water samples from Boulder Creek and the South Platte River were studied to test the method for two distinctly different natural-water matrices. The results from these analyses can be compared to those of the fortified reagent-water sample bias and variability data shown in tables 10 through 12.

In the Boulder Creek samples, the recovery of only one compound at one fortification, 1,7-dimethylxanthine at 0.10 μg/L, was greater than 100 percent. Excluding this result, compound-specific recoveries at both fortifications in Boulder Creek ranges between 22.1 and 80.1 percent. In the South Platte River samples, mean recoveries were much wider across all fortifications, with low recoveries of sulfamethoxazole and thiabendazole ranging between 14.1 and 30.5 percent, and high recoveries of 1,7-dimethylxanthine and albuterol ranging between 98.6 and 167 percent.

The differences in the range of recoveries for Boulder Creek samples when compared to the South Platte River samples provides additional insight into the specific effects of sample matrix upon pharmaceutical recovery. The absence of wastewater contributions to the sample matrix results in a markedly different distribution of pharmaceutical recoveries than is seen for South Platte River samples.

The closer convergence of overall mean and median recoveries in Boulder Creek samples when compared to South Platte samples suggests that the population of compound-specific mean recoveries was closer to a unimodal distribution than the more bimodal distribution of means observed for recoveries from South Platte River water. The overall lower and more uniformly distributed recoveries observed for Boulder Creek samples, when compared to the South Platte River samples, suggest that in the Boulder Creek samples, matrix enhancement and suppression play a smaller role and that the dominant effect of the sample matrix may be to compete for sorptive sites on the SPE stationary phase, resulting in lower overall recoveries. However, compound-specific matrix suppression (or enhancement) of the mass spectrometer signal may be important in other sample matrices that contain compositionally distinct or substantially higher DOC concentrations.

The high recoveries observed for 1,7-dimethylxanthene, and corresponding low recoveries for sulfamethoxazole and thiabendazole from the South Platte River samples, result from compound-specific matrix enhancement or suppression. At the South Platte River at Denver site where these samples were collected, the majority of the flow in the South Platte is typically derived from effluent discharge, and DOC concentrations typically are twice the DOC concentration in Boulder Creek. Evaluation of the average internal standard response and DOC concentration for each matrix indicates that some matrix suppression of the internal standard response occurs, despite testing and evaluation of multiple internal standards for minimal suppression. However, the average instrument response of each pharmaceutical between different sample matrices indicates that compound-specific matrix enhancements or suppressions cannot be corrected for by using average matrix response.

Thus, the analysis of matrix-spike recovery quality-control samples that are specific to the environmental water samples under study are required to accurately assess the effects of sample matrix on quantitation of pharmaceuticals in ambient environmental samples. In the case of the South Platte River samples, the matrix-spike recoveries suggest that the measured ambient environmental concentrations in these or similar samples may be over- or underestimates of the true concentration. Thus any study that quantifies concentrations of pharmaceuticals in environmental samples requires the routine inclusion of multiple matrix-spike recovery samples collected from the primary water types in the study area. Additional matrix-spike samples are needed if there are seasonal or other long-term variations in the study matrices. The matrix-spike sample results enable the evaluation of the specific effects of the water matrices in the study area using the quantitative results this analysis provides. The quantification of ambient environmental concentrations must include quantification of matrix effects.

Long-Term Measures of Variation of Analysis

To investigate more fully the overall effects of sample matrix upon compound recoveries, 38 pairs of environmental samples were collected across the United States over 2 years. One sample of each pair was fortified at 0.25 μg/L with method pharmaceuticals, and in most cases, the fortified samples were extracted and analyzed within the same sample set. The samples were collected as a part of the QA/QC of several research projects and included surface water, ground water, and treated effluent. Locations from which samples were acquired included Arizona, Florida, Kansas, New Jersey, North Dakota, and Oklahoma. Official USGS field methods applicable to sampling water for trace organic constituents (Wilde and others, 2004) were used for all sample collections. Samples were shipped by overnight express to the NWQL and were fortified at the NWQL just prior to analysis.

Recoveries of the method pharmaceuticals from the 38 fortified laboratory matrix-spike (LMS) samples, corrected for ambient aqueous pharmaceutical concentrations, are listed in table 22.

Overall, the mean and median of all pharmaceutical-specific mean recoveries was 75.3±30 and 78.0 percent, respectively. Pharmaceutical-specific, laboratory matrix-spike mean recoveries ranged between 36.0 and 117 percent. Individual sample-specific, matrix-spike recoveries were more variable, ranging between 0 (compound not detected because of matrix interferences) and 647 percent. However, examination of the 25th and 75th percentiles of recovery (table 22), as a measure of the central tendency of recoveries in this data set, suggests that, excluding high and low outliers, overall laboratory matrix-spike recoveries were acceptable, with overall median 25th and 75th percentiles of recovery, calculated from the data in table 22, of 56.8 and 91.0 percent, respectively. The maximum recoveries of specific pharmaceuticals of background-corrected laboratory matrix-spike samples ranged

Table 22. Recoveries of method pharmaceuticals determined in 38 laboratory matrix-spike samples and corrected for ambient aqueous pharmaceutical concentrations.

[All samples fortified at 0.25 microgram per liter. N, number of detections in 38 laboratory matrix-spike samples after correction for ambient aqueous pharmaceutical concentration]

Compound	N	Mean recovery (percent)	Standard deviation of recovery (percent)	Percent relative standard deviation of recovery (percent)	Median recovery (percent)	Minimum recovery (percent)	Maximum recovery (percent)	25th percentile of recovery (percent)	75th percentile of recovery (percent)
1,7-Dimethylxanthine	36	116	36.8	31.8	118	0.00	197	99.0	128
Acetaminophen	35	99.0	40.0	40.4	92.8	8.42	234	78.2	109
Albuterol	38	92.4	43.4	47.0	82.3	0.00	230	69.0	117
Caffeine	38	103	95.2	92.2	84.4	31.2	647	67.4	102
Carbamazepine	38	59.2	33.4	56.4	45.9	8.32	136	34.0	81.8
Codeine	38	117	60.8	52.0	100	7.24	283	85.9	118
Cotinine	38	86.5	32.3	37.4	82.0	41.0	217	67.9	92.2
Dehydronifedipine	38	99.0	34.8	35.1	96.0	14.0	204	80.8	111
Diltiazem	37	36.0	28.0	77.8	28.8	0.00	105	14.2	49.0
Diphenhydramine	37	41.6	32.4	77.8	32.2	0.00	149	17.0	55.4
Sulfamethoxazole	38	42.6	40.1	94.0	22.2	0.00	119	10.7	84.5
Thiabendazole	38	39.3	36.2	92.2	19.8	0.00	100	6.17	77.6
Trimethoprim	38	52.8	31.0	58.6	44.7	0.00	125	28.3	81.1
Warfarin	38	69.5	41.1	59.1	67.3	0.00	191	46.2	89.7

between 100 and 647 percent; however, the mean and median of those pharmaceutical-specific maximum recoveries were 210 and 194 percent, respectively. These recoveries range from acceptable to well outside the acceptable range.

Median recoveries for 1,7-dimethylxanthine, which displayed matrix enhancement in the South Platte River water samples, also were enhanced in the 38 laboratory matrix-spike samples (table 22). Median recoveries of carbamazepine, diltiazem, diphenhydramine, sulfamethoxazole, and thibendazole were less than 50 percent in this sample set, whereas in the South Platte River and Boulder Creek samples, sulfamethoxazole and thiabendazole were consistently recovered at less than 50 percent.

The causes of enhanced or suppressed recoveries for these compounds are difficult to assess in this data set, in the absence of comprehensive organic and inorganic analyses of these water samples, such as DOC concentrations or concentrations of contaminants from upstream contributions. In particular, the fact that these are single matrix samples fortified at a single concentration, corrected for ambient environmental concentrations using a single sample result analyzed by multiple operators and instruments adds variation and complicates comparison to the South Platte River and Boulder Creek data sets. Nevertheless, some conclusions can be drawn. The results suggest that in most cases, matrix enhancement affected the accuracy of quantitative results by about a factor of two or less, with occasional matrix enhancement as high as a factor of six. In particular, 1,7-dimethylxanthine, acetominophen, albuterol, codeine, caffeine, cotinine, and dehydronifedepine had maximum recoveries approaching, or greater than, 200 percent, and therefore, for environmental samples, caution is needed when comparing results for these compounds in samples from different sources.

Another potential source of uncertainty inherent in these results arises from the necessity of fortifying each LMS sample without knowing the ambient aqueous pharmaceutical concentrations, if any, of that sample. The analysis of the unfortified samples provides pharmaceutical concentrations, if present, that can correct recoveries for ambient environmental contributions, but the inherent uncertainty of the analytical measurement can result in either under- or overcorrection, particularly if the ambient environmental concentration is of similar or greater magnitude as the fortified concentration.

The relative contributions of ambient environmental concentrations to the total (that is, ambient plus fortified) concentrations of method pharmaceuticals were assessed in the 38 LMS-environmental sample pairs. Twelve of 14 method pharmaceuticals were detected in the 38 ambient environmental samples. Thiabendazole and warfarin were not detected in any of the ambient environmental samples. Ambient environmental concentrations of individual pharmaceuticals were detected in 4 to 20 samples, or in 10.3 to 53.9 percent of the samples, respectively. The most commonly detected pharmaceuticals were carbamazepine, cotinine, caffeine, and sulfamethoxazole, at frequencies of detection of 53.8, 51.3, 38.5, and 28.2 percent, respectively. The remaining pharmaceuticals were present in less than 25 percent of the samples. The respective median contributions of the ambient concentration to the total concentrations of these most commonly detected pharmaceuticals were 16.2, 3.23, 13.6, and 3.1 percent, respectively.

The pharmaceuticals whose ambient environmental concentrations contributed most to the total in the LMS samples were not the most frequently detected compounds, however. The pharmaceuticals with the highest median contributions of the ambient environmental concentrations to the total concentration in the LMS samples were 1,7-dimethylxanthine, acetominophen, trimethoprim, diphenhydramine, sulfamethoxazole, and diltiazem, at 65.6, 54.6, 47.0, 34.4, 33.9, and 27.6 percent, respectively. However, their respective frequencies of detection were 10.3, 18.0, 12.8, 12.8, 28.2, and 15.4 percent.

Wastewater LMS samples typically showed the largest contributions of ambient pharmaceuticals to the total LMS sample concentration; in two cases the ambient amount exceeded the fortified amount by a factor of 20 or greater.

These results suggest that ambient environmental concentrations do contribute to the LMS samples, in some cases substantially. Ambient environmental contributions to LMS samples from carbamazepine, cotinine, caffeine, and sulfamethoxazole are possible in most sample types, whereas the pharmaceuticals 1,7-dimethylxanthine, acetominophen, trimethoprim, diphenhydramine, sulfamethoxazole, and diltiazem, although less frequently detected, may contribute more to the total concentration in LMS samples, particularly wastewater. These results reinforce the need to carefully evaluate LMS and ambient environmental sample pairs as part of any study design for assessing pharmaceutical presence and concentration.

Long-Term Method Performance: Multiple Operators and Instruments

Long-term method performance under multiple operators and using multiple instruments, as reflected in statistical analysis of LRS samples over an extended time, provides a means of assessing the quality of pharmaceutical results for environmental samples collected over a similar period. For this method, the results of this statistical analysis also are used to evaluate whether concentrations of any given pharmaceutical should be routinely reported without specific qualification, or if the concentration should be reported as an estimated concentration because of less than optimal mean recovery or variation in recovery, as reflected in the standard deviation of recovery. This approach has been used previously for evaluating the reporting of polar pesticides determined by HPLC/MS (Furlong and others, 2001).

Results for 157 LRS samples extracted and analyzed between May 3, 2005, and May 4, 2006, are compiled and listed in table 23. This data period was used because it represents stable method performance after the method had been implemented at the NWQL as a routine research method and the analysts gained experience in its use. Mean recoveries ranged between 37.4 and 91.9 percent, with a similar range of median recoveries, 39.2 to 93.2 percent. The standard deviation of recovery for method pharmaceuticals in these LRS samples ranged between 8.41 and 21.2 percent. The limited range in standard deviations suggests relatively consistent performance in the absence of matrix effects during analysis of LRS samples under multiple operators and using multiple instruments.

Two statistics were used to determine if the reported concentration of any compound required quantitation qualification. Median recoveries calculated from long-term LRS data in table 23 are used to estimate the accuracy of recovered concentration. A nonparametric statistic, f-pseudosigma (Hoaglin, 1983), is calculated for the same data to determine the variation of LRS recoveries. The f-pseudosigma statistic is calculated from interquartile range (75th percentile minus the 25th percentile) of the data distribution divided by 1.349. The median and f-pseudosigma are used instead of the mean and standard deviation, respectively, because they minimize the effects of outlier values, and thus are more representative of long-term method performance under multiple operators and using multiple instruments.

Criteria for these measures were applied to the results in table 23 to evaluate whether concentrations for any of the pharmaceuticals should be routinely reported as estimates. Furlong and others (2001) previously had applied this approach as a means of including all aspects of the analytical process when routinely operated in a large-scale laboratory environment. Median recoveries had to fall within 60 and 120 percent for a compound to be considered reportable without qualification. Also, the f-pseudosigma statistic had to be less than 25 percent for a compound to be reported without qualification. Median recoveries and the f-pseudosigma statistics for those recoveries were determined for LRS samples extracted and analyzed between May 3, 2005, and May 4, 2006. The results of these calculations also are listed in table 23. The median recovery criteria indicate that 12 of 14 method pharmaceuticals are reported without qualification. The median recoveries for these 12 pharmaceuticals ranged between 61.3 and 93.2 percent. Concentrations of two compounds, diltiazem and warfarin, with median recoveries of 39.2 and 52.7 percent, respectively, are reported as estimates because their median recoveries did not meet the recovery criteria. All method pharmaceuticals met the f-pseudosigma criteria, with values ranging between 6.75 and 16.7.

These long-term recoveries are similar to the recoveries observed at the 0.05- and 0.10-µg/L reagent-water fortifications, and likely are more representative of overall method performance for pharmaceuticals in the absence of a natural sample matrix than the recoveries from the 10 organic-free water samples fortified at 0.25 µg/L and discussed in "Validation Results."

Comparison of Validation Sample and Long-Term Performance Sample Results

To evaluate the effect of sample matrix on pharmaceutical recovery and precision between different matrices, and to compare these results with long-term method performance, the median percent recovery, the standard deviation of recovery, and the f-pseudosigma of recovery for the 38 LMS samples (table 22) and 157 LRS samples (table 23) were combined with the median percent recovery, the standard deviation of recovery, and the f-pseudosigma of recovery calculated for four validation matrices for the individual validation sample results in tables 10 through 15 and 17 through 21. Note that

Table 23.   Recoveries of method pharmaceuticals determined in 157 laboratory reagent-spike samples extracted and analyzed over a 1-year period.

[All samples fortified at 0.25 microgram per liter. N, number of determinations]

Compound	N	Mean recovery (percent)	Standard deviation of recovery (percent)	Percent relative standard deviation of recovery (percent)	Median recovery (percent)	Minimum recovery (percent)	Maximum recovery (percent)	25th percentile of recovery (percent)	75th percentile of recovery (percent)	f-pseudosigma of recovery (percent)	Lower control limit (percent)	Upper control limit (percent)
1,7-Dimethylxanthine	157	83.7	21.2	25.3	79.7	49.3	146	71.7	88.3	12.3	42.8	117
Acetaminophen	157	67.1	14.2	21.1	66.6	34.1	108	59.1	77.7	13.8	25.3	108
Albuterol	150	70.2	14.2	20.2	72.0	35.0	103	61.2	81.0	14.7	28.0	116
Caffeine	157	91.9	15.9	17.3	93.2	57.3	144	79.3	101	16.2	44.7	142
Carbamazepine	157	85.0	9.80	11.5	87.0	54.9	110	79.1	91.4	9.15	59.6	114
Codeine	157	76.7	11.1	14.5	78.2	50.4	96.0	68.0	85.7	13.1	38.9	118
Cotinine	157	91.6	11.3	12.4	93.0	61.3	124	83.1	100	12.6	55.1	131
Dehydronifedipine	157	78.4	11.8	15.1	79.1	46.8	104	70.0	86.2	12.0	43.1	115
Diltiazem	157	37.4	15.0	40.1	39.2	5.48	72.8	26.3	48.8	16.7	-10.8	89.2
Diphenhydramine	157	59.5	8.41	14.1	61.3	32.0	83.9	55.2	64.3	6.75	41.1	81.5
Sulfamethoxazole	157	74.2	11.5	15.5	75.9	38.6	102	67.0	82.1	11.2	42.3	110
Thiabendazole	156	82.7	10.8	13.0	84.4	46.0	112	74.2	89.9	11.6	49.6	119
Trimethoprim	156	86.2	10.3	11.9	88.6	61.3	108	79.2	93.6	10.6	56.7	121
Warfarin	157	52.9	16.8	31.7	52.7	17.4	88.2	40.1	62.1	16.3	3.69	102
Surrogates[1]												
Carbamazepine-d_{10}	94	98.0	7.48	7.63	97.5	79.6	120	92.2	103	8.12	73.1	122
Ethyl nicotinate-d_4	157	85.5	12.7	14.9	86.2	50.9	116	78.9	94.2	11.3	52.2	120

[1]Surrogate compounds fortified at 0.5 microgram per liter.

the recovery results for each sample at each fortification concentration were combined into a single data set for each validation matrix, and the statistical tests described above were performed on this combined data set for each validation matrix. This combined data summary is listed in table 24.

It is critical that the assumptions implicit in summarizing the data in table 24 be understood to properly compare between the LMS, LRS, and validation matrix results in table 24. The LMS and LRS samples are fortified at a single concentration, and the fortified samples were analyzed over extended periods: 2 years and 1 year for the LMS and LRS samples, respectively. Each of these data sets can be considered a single homogeneous fortification experiment designed to evaluate method performance under multiple operator and instrument conditions, and the standard deviations and f-pseudosigmas of recovery for each pharmaceutical are expected to be larger, because they reflect more sources of variation than the validation matrix data sets, each of which combines fortifications at two or three concentrations, and the results for each concentration are from a single operator, single instrument experiment. If it is assumed that above the MDL, the variation of recovery, as reflected in the standard deviation of recovery, increases with concentration [an assumption common to chemical analysis and implicit in the calculation of the MDL (Oblinger Childress and others, 1999)], then the variation associated with the three fortifications of each validation matrix would increase with concentration. As a result, summarizing all fortifications of each matrix into a single data set combines the results of three distinct homogeneous experiments, and the aggregate standard deviations and f-pseudosigmas of recovery, as indicators of variation, may be larger than each individual fortification experiment.

The greater variation inherent in the combined fortification data sets of each validation matrix, compared to the individual fortifications, is apparent when the standard deviations of each validation matrix in table 24 are compared to the individual fortification results for each matrix in tables 10 through 15 and 17 through 21, where the aggregate standard deviation in table 24 is larger than the comparable single fortification experiments. As expected, the medians for all fortifications reported in table 24 are intermediate between the range of medians reported for the individual fortifications in tables 10 through 15 and 17 through 21. The f-pseudosigmas of recovery of the combined fortifications for each validation matrix are reported for comparison to the LMS and LRS results, and examination of table 24 suggests that the greatest variation of recovery is observed in the LMS results, compared to the ground-water, South Platte, and Boulder Creek samples, which was expected because the LMS results reflect matrix contributions from 38 different water samples, in addition to multiple operator and multiple instrument variation. Similarly, the matrix-free LRS samples have greater overall f-pseudosigmas of recovery than the reagent-water validation samples, a reflection of the multiple operator and multiple instrument sources of variation inherent in this data set.

It is difficult to identify a consistent pattern attributable to sample matrix when examining the median recoveries for individual pharmaceuticals in the six matrices in table 24.

For example 1,7-dimethyxanthine recoveries appear to be consistently higher in LMS, ground-water, South Platte, and Boulder Creek samples, when compared to the matrix-free LRS and reagent-water validation samples. However, median cotinine recoveries are variable between the six matrices, with the highest median recovery, 105 percent, in the South Platte surface-water samples; the lowest median recovery, 53.5 percent, in the Boulder Creek surface-water samples; and median recoveries for the matrix-containing LMS and ground-water samples, and matrix-free LRS and reagent-water samples falling in between the surface-water samples.

To better assess method performance differences between the the LMS, LRS, and the four validation sample types, the grand mean, grand median, and grand standard deviation were calculated from the median recoveries, standard deviations of recovery, and f-pseudosigmas of recovery for all pharmaceuticals determined in the LMS, LRS, and the four validation sample types reported in table 24. The grand means, medians, and standard deviations for all six sample types are listed in table 25 and were calculated from the summary results for individual pharmaceuticals for each sample type in table 24.

The data listed in table 25 condense method performance for each sample type into a small set of statistics that aggregate individual pharmaceutical performance for each sample type, and the grand mean (GMN), grand median (GMD), and grand standard deviation (GSD), when cautiously interpreted, can provide useful comparisons of overall method performance between each sample type. Similar GMN and GMD values for the median, standard deviation, and f-pseudosigma of recovery in each sample type suggest that the individual pharmaceutical medians, standard deviations, and f-pseudosigmas of recovery aggregated in the GMN, GMD, and GSD results approximate a unimodal distribution and may be compared between sample types. The difference between GMNs and GMDs for medians, standard deviations, and f-pseudosigmas of recovery for all six sample types is less than 6 percent, and more typically is between 2 and 3 percent. The GSD of median recoveries in a sample type reflects the range of individual pharmaceutical median recoveries in that sample type, and a small GSD suggests a narrower, more uniform range of median recoveries.

The GMDs of median recoveries ranged from 50.8 percent for Boulder Creek samples to 86.8 percent for ground-water samples. The reagent water GMD, 84.5 percent, was similar to ground-water GMD, suggesting comparable performance in these two validation sample types. The GMD of the LRS samples is 78.6 percent. A similar range of median recoveries is reflected in the GSDs for the LRS, the reagent-water validation, and ground-water validation samples, which are 15.6, 18.4, and 16.9 percent, respectively. The GSDs for the LMS, South Platte, and Boulder Creek samples are 26.4, 41.8, and 22.3 percent, respectively; these somewhat greater GSDs may reflect the wider range of pharmaceutical-specific matrix enhancements or suppressions discussed earlier for these sample types, and are likely reflective of the greater sample matrix contributions in these sample types compared to the matrix-free LRS and reagent-water samples, and the relatively low concentration of matrix, as reflected by DOC, in the ground-water samples.

Table 24. Median, standard deviation, and f-pseudosigma of recovery for laboratory matrix-spike samples, laboratory reagent-water spike samples, and reagent-water, ground-water, South Platte surface-water, and Boulder Creek surface-water validation samples, combining all results for each fortification in each matrix.

Compound name	Laboratory matrix spikes—38 observations from one fortification concentration			Laboratory reagent-water spikes—157 observations from one fortification concentration			Reagent-water validation samples—28 observations from three fortification concentrations		
	Median recovery (percent)	Standard deviation (percent)	f-Pseudosigma of recovery (percent)	Median recovery (percent)	Standard deviation (percent)	f-Pseudosigma of recovery (percent)	Median recovery (percent)	Standard deviation (percent)	f-Pseudosigma of recovery (percent)
1,7-Dimethylxanthine	119	36.8	21.2	79.7	25.3	12.3	102	15.4	13.5
Acetaminophen	98.2	40.0	23.0	66.6	21.1	13.8	78.8	24.6	28.4
Albuterol	82.3	43.4	35.8	72.0	20.2	14.7	69.0	25.0	35.0
Caffeine	86.7	95.2	26.0	93.2	17.3	16.2	108	12.6	13.7
Carbamazepine	55.9	33.4	35.4	87.0	11.5	9.15	95.0	10.8	12.0
Codeine	102	60.8	23.7	78.2	14.5	13.1	81.1	11.0	11.8
Cotinine	82.5	32.3	18.1	93.0	12.4	12.6	80.0	13.6	14.9
Dehydronifedipine	101	34.8	22.1	79.1	15.1	12.0	106	11.3	8.72
Diltiazem	31.9	28.0	25.8	39.2	40.1	16.7	57.0	7.76	7.41
Diphenhydramine	38.9	32.4	28.4	61.3	14.1	6.75	55.3	8.11	7.19
Sulfamethoxazole	52.2	40.1	54.7	75.9	15.5	11.2	112	11.2	9.33
Thiabendazole	48.2	36.2	53.0	84.4	13.0	11.6	82.9	8.48	10.3
Trimethoprim	60.2	31.0	39.1	88.6	11.9	10.6	86.2	8.18	8.90
Warfarin	73.6	41.1	32.3	52.7	31.7	16.3	104	13.8	11.2

Compound name	Ground-water validation samples—28 to 29 observations from three fortification concentrations			South Platte River surface-water validation samples—29 to 30 observations from three fortification concentrations			Boulder Creek surface-water validation samples—19 to 20 observations from two fortification concentrations		
	Median recovery (percent)	Standard deviation (percent)	f-Pseudosigma of recovery (percent)	Median recovery (percent)	Standard deviation (percent)	f-Pseudosigma of recovery (percent)	Median recovery (percent)	Standard deviation (percent)	f-Pseudosigma of recovery (percent)
1,7-Dimethylxanthine	111	20.4	27.6	150	21.0	14.1	117	29.8	41.9
Acetaminophen	108	9.26	12.4	111	20.8	10.5	74.2	10.4	12.3
Albuterol	88.4	12.3	13.1	138	22.6	26.0	68.9	8.59	11.7
Caffeine	90.3	18.8	24.8	77.9	17.3	11.1	51.2	8.45	8.35
Carbamazepine	67.5	11.9	14.5	58.9	9.34	13.3	46.3	5.52	6.86
Codeine	85.2	6.97	6.38	121	16.8	17.8	58.0	7.44	10.2
Cotinine	77.9	18.0	21.7	105	17.2	21.1	53.5	8.21	7.64
Dehydronifedipine	99.2	6.55	3.93	119	14.5	13.0	67.8	10.1	12.5
Diltiazem	60.2	4.74	3.71	51.9	5.13	4.11	41.3	6.97	9.35
Diphenhydramine	62.8	3.27	3.48	51.4	4.61	3.99	39.9	7.07	9.77
Sulfamethoxazole	96.6	10.3	13.0	18.4	9.30	8.49	30.0	7.19	9.65
Thiabendazole	61.2	5.74	4.97	23.9	8.84	7.15	29.6	7.32	8.38
Trimethoprim	77.2	5.54	5.13	52.4	6.39	7.74	45.2	6.92	9.39
Warfarin	91.0	11.8	15.2	81.7	6.30	5.87	50.5	7.66	10.6

Table 25. Grand means, medians, and standard deviations of the median, standard deviation, and *f*-pseudosigma of recovery of all pharmaceuticals in each matrix (laboratory matrix-spike samples, laboratory reagent-water spike samples, and reagent-water, ground-water, South Platte surface-water, and Boulder Creek surface-water validation samples) listed in table 24.

	Laboratory matrix spikes—38 observations from one fortification concentration			Laboratory reagent-water spikes—157 observations from one fortification concentration			Reagent-water validation samples—28 observations from three fortification concentrations		
	Median recovery (percent)	Standard deviation (percent)	*f*-Pseudosigma of recovery (percent)	Median recovery (percent)	Standard deviation (percent)	*f*-Pseudosigma of recovery (percent)	Median recovery (percent)	Standard deviation (percent)	*f*-Pseudosigma of recovery (percent)
Grand mean	73.7	41.8	31.3	75.1	18.8	12.6	86.9	13.0	13.7
Grand median	78.0	36.5	27.2	78.6	15.3	12.4	84.5	11.3	11.5
Grand standard deviation	26.4	17.3	11.3	15.6	8.4	2.82	18.4	5.51	8.07

	Ground-water validation samples—28 to 29 observations from three fortification concentrations			South Platte River surface-water validation samples—29 to 30 observations from three fortification concentrations			Boulder Creek surface-water validation samples—19 to 20 observations from two fortification concentrations		
	Median recovery (percent)	Standard deviation (percent)	*f*-Pseudosigma of recovery (percent)	Median recovery (percent)	Standard deviation (percent)	*f*-Pseudosigma of recovery (percent)	Median recovery (percent)	Standard deviation (percent)	*f*-Pseudosigma of recovery (percent)
Grand mean	84.0	10.4	12.1	82.9	12.9	11.7	55.2	9.40	12.0
Grand median	86.8	9.78	12.7	79.8	11.9	10.8	50.8	7.55	9.71
Grand standard deviation	16.9	5.50	8.12	41.8	6.41	6.45	22.3	6.01	8.74

It is important to recognize that precision of results is less affected by the sample matrix than the concentration or recovery. This is reflected in the GMDs of the standard deviations and f-pseudosigmas of recovery for each matrix type (table 25). The GMD of the standard deviation ranged between 7.55 percent in Boulder Creek samples and 36.5 percent in the LMS samples; if the LMS GMD is excluded, the highest GMD of the standard deviation is 15.3 percent for the LRS samples. The GMD of the f-pseudosigmas of recovery ranges between 9.71 percent in Boulder Creek samples and 27.2 percent in the LMS samples; if the LMS GMD is excluded, the highest GMD of the f-pseudosigma of recovery is 12.7 percent for the ground-water samples. Thus, with the exception of the LMS samples, the typical median recovery and f-pseudosigma of recovery for a pharmaceutical in all matrices falls well within the 25-percent tolerance used previously by Furlong and others (2001) to define acceptable variation in large (150 samples or more) data sets.

With the exception of the LMS samples, the data in table 25 suggest that results produced using this method are sufficiently precise to make comparisons between samples with similar matrices, such as between samples in a study within a watershed. Note that comparison of pharmaceutical concentration differences between samples from different water sources or with substantially different matrix concentrations requires complementary matrix-spike sample data to ensure that absolute differences between two samples with different matrices are not an artifact of differing levels of matrix enhancement or suppression.

Blank Contamination Study

The potential for blank contamination in this method is of particular concern because the method pharmaceuticals are commonly used as over-the-counter and prescription pharmaceutical formulations. Cotinine, a nicotine metabolite, and caffeine are components in commonly used consumer products, or are degradation products of those components. As a result, the potential for sample contamination during analysis must be carefully assessed. A set of LRB samples, analyzed during stable method performance and corresponding to the same time interval as the long-term LRS sample set discussed in the "Long-Term Measures of Variation of Analysis" section, was evaluated to assess whether blank contamination during sample handling, extraction, or analysis could occur routinely. The 1-year long data set enables an evaluation of episodic and chronic contamination.

The frequency of detection and observed concentrations of method pharmaceuticals in 99 LRB samples are listed in table 26. Eleven of 14 pharmaceuticals measured in this method were detected in one or more LRB samples, and 9 out of 14 pharmaceuticals were detected at maximum concentrations greater than 0.005 µg/L. Excluding diphenhydramine,

(discussed further below) and the three compounds not detected in any LRB samples, the frequency of detections ranged from 1.0 to 5.1 percent, which corresponds to 1 to 5 detections in 99 blanks. The overall median frequency of detection of any method pharmaceutical in any blank was 3 percent, or 3 detections in 99 samples. With the exception of diphenhydramine, detections were scattered across the time period encompassed by the blank samples. Forty-four method pharmaceuticals were detected, of a maximum 1,386 possible detections in 99 LRB samples. Of these, 18 detections were present as isolated detections, and there were seven sequential pairs of detections; four of the sequential pairs of detections occurred in the same two LRB samples, suggesting that overall systemic or chronic contamination was not substantial during this period. One pharmaceutical, diphenhydramine, was detected in 14 of 99 LRB samples, a frequency initially suggesting chronic contamination. However, evaluation of the blanks indicated that 12 of the 14 blank detections of diphenhydramine occurred in a sequential set of blanks, reflecting episodic blank contamination that was corrected. The mean concentration of diphenhydramine in the LRB samples was 0.0054 µg/L. Some compounds, such as caffeine and codeine, were less frequently detected but when detected, were at appreciable concentrations. Caffeine, detected in 5 of 99 LRB samples, was present at a mean concentration of 0.0778 µg/L, and at a maximum concentation of 0.239 µg/L, exceeding the method detection limit (MDL; table 27) of 0.0075 µg/L. Codeine, detected in 3 of 99 samples, was present at a mean concentration of 0.0150 µg/L, and at a maximum of 0.0334 µg/L, exceeding the MDL of 0.0111 µg/L. To ensure that inadvertent contamination during sample extraction, isolation, and analysis is not reported as environmental concentrations, a consistent approach to qualifying data, such as that documented in Maloney (2005), is needed to carefully evaluate LRB samples.

In summary, contamination by method pharmaceuticals during sample processing and analysis is infrequent, and set-specific LRB sample results are sufficient to evaluate environmental sample results for the potential episodic presence of method pharmaceuticals that could be inadvertently introduced in the laboratory during sample processing, extraction, and instrumental analysis. However, routine evaluation of LRB data sets is useful in identifying compound-specific or sequential sample-specific contamination and is needed for long-term, routine application of this method. Similarly, because the pharmaceuticals determined in this method are commonly used, these results also emphasize the need for field blank sample collection to ensure that environmental concentrations of pharmaceuticals do not result from inadvertent introduction during sample collection and processing. These field blank samples, like the LMS and replicate samples discussed earlier, need to be a part of the project quality-assurance plan.

Table 26. Detections of method pharmaceuticals determined in 99 laboratory reagent-blank samples extracted and analyzed over a 1-year period.

[ND, not detected; NA, not applicable]

Compound	Mean of detected concentrations, in micrograms per liter	Maximum concentration detected, in micrograms per liter	Number of detections, of 99 possible	Frequency of detection, in percent
1,7-Dimethylxanthine	0.0612	0.0612	1	1.0
Acetaminophen	.0166	.0374	3	3.0
Albuterol	ND	NA	0	NA
Caffeine	.0778	.239	5	5.1
Carbamazepine	.0002	.0002	1	1.0
Codeine	.0150	.0334	3	3.0
Cotinine	.0074	.0136	3	3.0
Dehydronifedipine	.0001	.0002	4	4.0
Diltiazem	.0037	.0067	4	4.0
Diphenhydramine	.0054	.0191	14	14
Sulfamethoxazole	ND	NA	0	NA
Thiabendazole	.0103	.0103	1	1.0
Trimethoprim	.0024	.0067	5	5.1
Warfarin	ND	NA	0	NA

Table 27. Method detection limits and interim reporting levels calculated from eight replicate determinations of method pharmaceuticals fortified in organic-free reagent water at 0.05 microgram per liter.

Compound	Mean recovery, in micrograms per liter	Standard deviation of recovery, in micrograms per liter	Method detection limit, in micrograms per liter	Interim reporting level, in micrograms per liter
1,7-Dimethylxanthine	0.0585	0.0035	0.0104	0.020
Acetaminophen	.0591	.0040	.0119	.025
Albuterol	.0570	.0023	.0069	.015
Caffeine	.0576	.0025	.0075	.015
Carbamazepine	.0482	.0030	.0089	.030
Codeine	.0494	.0037	.0111	.020
Cotinine	.0465	.0047	.0142	.030
Dehydronifedipine	.0569	.0037	.0110	.020
Diltiazem[*]	.0341	.0030	.0089	.040
Diphenhydramine	.0298	.0038	.0115	.050
Sulfamethoxazole	.0614	.0040	.0119	.10
Thiabendazole	.0418	.0042	.0125	.10
Trimethoprim	.0476	.0034	.0102	.040
Warfarin[*]	.0570	.0031	.0094	.020

[*]Routinely reported as an estimated concentration, indicated by an "E" qualifier.

Reporting Limits

Method detection limits (MDLs) were determined for this method by using the procedures of the U.S. Environmental Protection Agency (2005). The MDL is defined as the minimum concentration of a substance that can be measured and reported with 99-percent confidence that the compound concentration is greater than zero. Initial MDLs were determined from eight replicate analyses of reagent-water samples of 1-L volume, fortified to 0.05 µg/L.

The MDL was calculated according to equation 7:

$$MDL = S \times t_{(n-1,\ 1-\alpha = 0.99)} \tag{7}$$

where

S =   standard deviation of replicate analyses, in micrograms per liter;

n =   number of replicate analyses;

and

$t_{(n-1,\ 1-\alpha = 0.99)}$ =   Student's t-value for the 99-percent confidence level with $n-1$ degrees of freedom.

According to the U.S. Environmental Protection Agency (USEPA) procedure, at least seven replicate samples are fortified with compounds at concentrations of two to five times the estimated MDL. Data for the MDL determination in this study were taken from the lowest concentration fortifications used to determine method recoveries, 0.05 µg/L for organic-free reagent water. The calculated MDLs are listed in table 27, as are the interim reporting levels (IRLs), which are at least two times the MDL. The IRL is used by the NWQL because it reduces the risk of reporting false positives (Oblinger Childress and others, 1999). Some pharmaceutical IRLs were adjusted to greater than two times the MDL to reflect consistently lower recovery.

The MDLs for method pharmaceuticals ranged from 0.0069 µg/L for albuterol to 0.0142 µg/L for cotinine, with a median overall MDL for all pharmaceuticals of 0.0107 µg/L. These MDLs were slightly lower than expected from the MDL estimates that were used to determine the spiking concentration, but are consistent with the MDLs reported by Cahill and others (2004) for an earlier version of this method and indicate consistency during the transition of this method from a research to production setting. Interim reporting levels ranged between 0.015 and 0.10 µg/L. A program of long-term method detection level determination (Oblinger Childress and others, 1999), described below, is used to assess the need for adjusting the MDL and IRL of each pharmaceutical as the method is used at the NWQL.

Qualitatively identified compounds whose calculated concentrations are less than the IRL are reported as estimated and noted with the "E" remark code because this method is an "information-rich" method, as are other MS methods used by the USGS (Oblinger Childress and others, 1999). As Oblinger Chlidress and others (1999) note, reporting estimated concentrations provides a richer data set that may be explored to better understand environmental distributions, as long as all qualitative identification criteria used for mass spectral methods are met. Estimated concentrations also provide important information about environmental concentrations that can be

used to improve analytical methods. Pharmaceutical concentrations less than 0.003 µg/L are not reported because extensive experience demonstrated that at this concentration, typical instrument responses for most pharmaceuticals cannot be reliably distinguished from background instrument noise.

As part of normal quality assurance/quality control practices of the Office of Water Quality, U.S. Geological Survey, the MDLs and IRLs for method pharmaceuticals are evaluated over an extended period (6 to 12 months) and include MDL and IRL determinations from a sufficient number ($n > 30$) of samples to reflect multiple instruments, analysts, and calibration curves. These long-term method detection levels (LT–MDL; Oblinger Childress and others, 1999) provide a more accurate assessment of method performance under the conditions in which samples for long-term monitoring programs are analyzed. The MDLs and IRLs provide an initial point of reference for evaluating changes in method performance when compared to the LT–MDL. The process used to evaluate LT–MDLs and to make changes in MDLs and IRLs is described more fully at the "Long-Term Method Detection Levels" web page maintained by the Branch of Quality Systems, Office of Water Quality, U.S. Geological Survey at URL *http://bqs.usgs.gov/ltmdl/* last accessed January 3, 2008. Changes in the pharmaceutical method LT–MDLs also are reported at this website.

Other Pharmaceuticals Evaluated for This Method

Eighteen additional pharmaceuticals were tested for inclusion in this method, but are not part of the final method; these pharmaceuticals are listed in table 28. Fourteen of these compounds responded sufficiently well under positive electrospray ionization conditions, but were insufficiently isolated on the SPE phase used in the method. Insufficient retention was defined as recoveries less than 35 percent from fortified reagent-water samples. Three compounds, amoxicillin, cephalexin, and urobilin, did not respond sufficiently under the positive electrospray conditions used in this method and were removed from consideration prior to SPE testing. One pharmaceutical, ibuprofen, included in Cahill and others (2004), was not included in the final list of pharmaceuticals in this method because of insufficient sensitivity under the positive electrospray conditions used in this method, although it met criteria for recovery from SPE.

Sample Holding-Time Study

Holding-time studies were conducted for water samples and water-sample extracts. Results from the water sample holding-time study indicate that samples should be stored at 4°C and extracted within 4 days of collection to ensure that sample results are minimally affected by degradation. The water-sample extract holding-time study showed that for most pharmaceuticals, acceptable results can be obtained from extracts stored at 4°C for up to 35 days, far longer than extracts are typically held in the HPLC/MS autosampler, even when sample sets are combined into larger analytical batches. The sample holding-time study is described in detail in Appendix B.

Table 28. Pharmaceutical compounds tested for inclusion in this method and the reason that each compound was not included.

[ESI, electrospray ionization]

Compound	Reason for failure in this method
Amoxicillin	Insufficient ionization under positive ESI
Azithromycin	Insufficient recovery from extraction, less than 30 percent
Cephalexin	Insufficient ionization under positive ESI
Cimetidine	Insufficient recovery from extraction, less than 30 percent
Clarithromycin	Insufficient recovery from extraction, less than 30 percent
Digoxigenin	Insufficient recovery from extraction, less than 30 percent
Digoxin	Insufficient recovery from extraction, less than 30 percent
Enalaprilat	Insufficient recovery from extraction, less than 30 percent
Erythromycin	Insufficient recovery from extraction, less than 30 percent
Fluoxetine	Insufficient recovery from extraction, less than 30 percent
Furosemide	Insufficient recovery from extraction, less than 30 percent
Gemfibrozil	Insufficient recovery from extraction, less than 30 percent
Ibuprofen	Insufficient ionization under positive ESI
Lisinopril	Insufficient recovery from extraction, less than 30 percent
Metformin	Insufficient recovery from extraction, less than 30 percent
Miconazole	Insufficient recovery from extraction, less than 30 percent
Ranitidine	Insufficient recovery from extraction, less than 30 percent
Urobilin	Insufficient ionization under positive ESI

Application of This Method to Environmental Studies

The development and application of this method has demonstrated the important effects sample matrix can play upon recovery of pharmaceuticals from natural-water samples. Additionally, the potentially confounding effect that natural water sample matrix can have upon ionization of pharmaceuticals has been demonstrated, either by enhancing or suppressing ionization. Sample fortification and recovery experiments in reagent-, ground-, and surface-water samples showed that individual pharmaceuticals may be over- or underestimated as a result of matrix effects. The sample matrices used in this study, particularly that of the surface-water samples, were chosen to reflect common water types, including surface water with substantial influence from wastewater effluent, but these matrices are not representative of all matrices to which this method may be applied. Thus laboratory matrix spikes are required for water-quality studies using this method to assess the presence and distribution of pharmaceuticals.

The selection of specific water types for laboratory matrix-spike samples is a critical aspect of project quality-control plans, and the study site should be carefully evaluated to determine the primary water types present, which will define the number of laboratory matrix spikes necessary. For example, if a study site consists of a river reach with wastewater-treatment-plant discharges to the river, at a minimum, two laboratory matrix-spike samples would be required, one upstream from the wastewater discharge, and one immediately downstream or in the discharge itself. If the study were to sample multiple times over an extended period, or if there are other wastewater-treatment plants or water sources within the study area, these also would require additional laboratory matrix-spike samples to describe the primary

water types in the study area over the duration of the study. The planning and collection of laboratory matrix spikes requires careful consideration to ensure that samples encompass the range of water types present in the study area and that they meet the data quality objectives of the study. Consultation with appropriate experts in the hydrology and water quality of the study area and the chemistry of pharmaceuticals is strongly encouraged.

Replicate environmental water samples and field equipment blank samples are a necessary complement to the laboratory matrix-spike samples, and their collection and analysis should be planned in conjunction with the collection of laboratory matrix-spike samples. The replicate field samples provide an estimate of the combined precision of the field sample collection and analytical method that cannot be adequately addressed with the precision data collected and analyzed in this report. Because many of the pharmaceuticals measured in this study are commonly used, the potential for contamination during sample collection and field processing must be assessed. Field equipment blank samples, collected and processed as outlined in Wilde and others (2004), and evaluated in conjunction with laboratory blank samples, ensure that low-level detection of pharmaceuticals in water samples do not result from contamination during sample collection and processing.

Summary and Conclusions

The U.S. Geological Survey (USGS) National Water Quality Laboratory has developed an analytical method for the determination of 14 pharmaceuticals in aqueous samples, including ground water, surface water, and domestic wastewater. This method uses solid-phase extraction coupled with high-performance liquid chromatography/mass spectrometry to sensitively and selectively detect these compounds at the expected ambient environmental concentrations, which typically range between 0.01 and 0.1 microgram per liter (µg/L). The extraction component of this method can be operated manually or by using automated solid-phase extraction instrumentation. This method provides an efficient means of detecting and quantifying important, pharmaceutically active compounds that typically might not be reported because they are unregulated or not included in other USGS, U.S. Environmental Protection Agency, American Water Works Association, or other official methods. In this method, the concentrations of 12 pharmaceuticals are reported without qualification; the concentrations of two pharmaceuticals are reported as estimates because long-term reagent-spike sample recoveries fall below acceptance criteria for reporting concentrations without qualification.

Water samples are collected and the pharmaceuticals of interest are isolated by solid-phase extraction with a modified styrene-divinylbenzene stationary phase and are determined by high-performance liquid chromatography/mass spectrometry using positive electrospray ionization operated in the selected-ion monitoring mode to reduce chemical noise and to improve specificity. The pharmaceuticals in this method are representative of a range of pharmacologically active compound classes that are reflective of contemporary prescribing and human-use patterns. Because human wastewater is an important source for these compounds, this method complements other official methods of the USGS that measure wastewater indicators in water, such as ethoxylate surfactants, fragrances, food additives, antioxidants, phosphate flame retardants, plasticizers, industrial solvents, disinfectants, and fecal sterols.

The single-operator standard deviation at 0.05 μg/L for individual pharmaceuticals in organic-free water samples ranged from 4.62 to 9.47 percent. Recoveries in organic-free water samples ranged from 59.5 to 123 percent. More variation in analyte recovery was observed in two surface-water samples, reflected in phamaceutical-specific mean recoveries ranging between 14.1 and 167 percent. These results reflect competing aspects of matrix (dissolved organic carbon) interferences, particularly matrix competition for the solid-phase extraction stationary phase, and matrix enhancement or suppression affecting electrospray ionization.

For the multiple operator, multiple instrument data set of 157 organic-free water samples fortified at 0.25 μg/L and analyzed for 1 year from May 3, 2005, to May 4, 2006, the mean relative standard deviation for all pharmaceuticals is 18.8 percent. Mean recoveries in these same samples averaged 75.1 percent for all compounds. The mean long-term set fortification recoveries of 12 of 14 pharmaceuticals in this method fell between acceptance limits of 60 and 120 percent, and concentrations of these compounds are reported without qualification. The other two compounds are reported as qualified estimates. Metaanalysis of laboratory matrix spike, laboratory reagent spike, and validation sample recovery data indicate that while matrix effects can affect absolute recovery, overall precision within a sample type were typically much less than 25 percent, facilitating comparison between samples with similar levels and types of matrix.

Nine out of 14 pharmaceuticals were sporadically detected in blanks at concentrations greater than 0.005 μg/L, but laboratory reagent-blank samples are sufficient to qualify the results of specific sample sets when this infrequent contamination occurs. The single operator, single instrument validation data and the long-term quality-control data reported here provide strong evidence for the application of high-performance liquid chromatography/mass spectrometry to large-scale, routine monitoring programs for pharmaceuticals in surface-, ground-, and wastewater for environmental concentrations at or less than 10 parts per trillion (0.010 μg/L or 10 ng/L).

Holding-time studies were conducted for water samples and water-sample extracts. Results from the water sample holding-time study indicate that samples should be stored at 4°C and extracted within 4 days of collection to ensure that sample results are minimally affected by degradation. The water-sample extract holding-time study showed that for most pharmaceuticals, acceptable results can be obtained

from extracts stored at 4°C for up to 35 days, far longer than extracts are typically held in the HPLC/MS autosampler, even when sample sets are combined into larger analytical batches.

The method detection limits and interim reporting levels for the pharmaceuticals determined by this method were calculated from recoveries of pharmaceuticals in reagent-water samples amended at 0.05 μg/L, and ranged between 0.0069 and 0.042 μg/L, and 0.015 and 0.10 μg/L, respectively. The concentrations of 12 pharmaceuticals are reported without qualification; the concentrations of two pharmaceuticals are reported as estimates because long-term reagent-spike sample recoveries fall below acceptance criteria for reporting concentrations without qualification. Pharmaceutical concentrations less than 0.003 μg/L are not reported because at this concentration, typical instrument responses for most pharmaceuticals cannot be reliably distinguished from background instrument noise.

This report documents the effects of sample matrix upon the quantitative results for some pharmaceuticals determined by this method. Thus laboratory matrix-spike samples collected from representative water types within a study are required to assess the water type-specific matrix effects upon the results from water samples collected to determine the presence and distribution of pharmaceuticals. These laboratory matrix-spike samples are collected and analyzed in addition to the replicate water samples and field equipment blank samples that are part of a study quality assurance/quality control plan.

References Cited

Barber, L.B., Brown, G.K., Cahill, J.D., Furlong, E.T., and Keefe, S.H., 2003, Natural and contaminant organic compounds in the Boulder Creek Watershed, Colorado, under high-flow and low-flow conditions, 2000 (chap. 5), *in* Murphy, S.F., Verplanck, P.L., and Barber, L.B., eds., Comprehensive water quality of the Boulder Creek Watershed, Colorado, during high-flow and low-flow conditions, 2000: U.S. Geological Survey Water-Resources Investigations Report 03–4045, p. 103–144.

Barber, L.B., Verplank, P.L., Murphy, S.F., Sandstrom, M.W., Taylor, H.E., and Furlong, E.T., 2006, Chemical loading into surface water along a hydrological, biogeochemical, and land use gradient: A holistic watershed approach: Environmental Science & Technology, v. 40, p. 475–486.

Bossong, C.R., Caine, J.S., Stannard, D.I., Flynn, J.L., Stevens, M.R., Heiny-Dash, J.S., 2003, Hydrologic conditions and assessment of water resources in the Turkey Creek watershed, Jefferson County, Colorado, 1998–2001: U.S. Geological Survey Water-Resources Investigations Report 03–4034, 140 p.

Cahill, J.D., 2000, Determination of human use pharmaceuticals in surface water using solid-phase extraction and high-performance liquid chromatography–mass spectrometry: M.S. Thesis submitted to the University of Colorado at Denver, 112 p.

Cahill, J.D., Furlong, E.T., Burkhardt, M.R., Kolpin, D.W., and Anderson, L.G., 2004, Determination of pharmaceutical compounds in surface- and ground-water samples by solid-phase extraction and high-performance liquid chromatography/electrospray ionization mass spectrometry: Journal of Chromatography A, v. 1041, p. 171–180.

Daughton, C.G., and Ternes, T.A., 1999, Pharmaceuticals and personal care products in the environment: Agents of subtle change? Environmental Health Perspectives, v. 107, p. 907–938.

Enke C.G., 1997, A predictive model for matrix and analyte effects in electrospray ionization of singly-charged ionic analytes: Analytical Chemistry, v. l69, no. 23, p. 4885–4893.

Furlong, E.T., Burkhardt, M.R., Gates, P.M., Werner, S.L., and Battaglin, W.A., 2000, Routine determination of sulfonylurea, imidazolinone, and sulfonamide herbicides at nanogram-per-liter concentrations by solid-phase extraction and liquid chromatography/mass spectrometry: The Science of the Total Environment, v. 248, nos. 2–3, p. 135–146.

Furlong, E.T., Anderson, B.D., Werner, S.L., Soliven, P.P., Coffey, L.J., and Burkhardt, M.R., 2001, Methods of analysis by the U.S. Geological Survey National Water Quality Laboratory—Determination of pesticides in water by graphitized carbon-based solid-phase extraction and high-performance liquid chromatography/mass spectrometry: U.S. Geological Survey Water-Resources Investigations Report 01–4134, 73 p.

Glassmeyer, S.T., Furlong, E.T., Kolpin, D.W., Cahill, J.D., Zaugg, S.D., Werner, S.L., Meyer, M.T., and Kryak, D.D., 2005, Transport of chemical and microbial compounds from known wastewater discharges: Potential for use as indicators of human fecal contamination: Environmental Science & Technology, v. 39, p. 5157–5169.

Hoaglin, D.C., 1983, Letter values—A set of selected order statistics, *in* Hoaglin, D.C., Mosteller, F., and Tukey, J.W., eds., Understanding robust and exploratory data analysis: New York, John Wiley and Sons, p. 33–57.

Kebarle, P., and Ho, Y., 1997, On the mechanism of electrospray mass spectrometry, *in* Cole, R.B., ed., Electrospray ionization mass spectrometry—Fundamentals, instrumentation and applications: New York, Wiley-Interscience, p. 3–63.

Kolpin, D.W., Furlong, E.T., Meyer, M.T., Thurman, E.M., Zaugg, S.D., Barber, L.B., and Buxton, H.T., 2002, Pharmaceuticals, hormones, and other organic wastewater contaminants in U.S. streams, 1999–2000—A national reconnaissance: Environmental Science & Technology, v. 36, no. 6, p. 1202–1211.

Litke, D.W., and Kimbrough, R.A., 1998, Water-quality assessment of the South Platte River basin, Colorado, Nebraska, and Wyoming—Environmental setting and water quality of fixed sites, 1993–95: U.S. Geological Survey Water-Resources Investigations Report 97–4220, 61 p.

Maloney, T.J., ed., 2005, Quality management system, U.S. Geological Survey National Water Quality Laboratory: U.S. Geological Survey Open-File Report 2005–1263, version 1.3, 9 November 2005, chapters and appendixes variously paged.

Murphy, S.F., Verplanck, P.L., and Barber, L.B., ed., 2003, Comprehensive water quality of the Boulder Creek Watershed, Colorado, during high-flow and low-flow conditions, 2000: U.S. Geological Survey Water-Resources Investigations Report 03–4045, 213 p.

Oblinger Childress, C.J., Foreman, W.T., Connor, B.F., and Maloney, T.J, 1999, New reporting procedures based on long-term method detection levels and some considerations for interpretations of water-quality data provided by the U.S. Geological Survey National Water Quality Laboratory: U.S. Geological Survey Open-File Report 99–193, 19 p.

Sando, S.K., Furlong, E.T., Gray, J.L., Meyer, M.T., and Bartholomay, R.C., 2005, Occurrence of organic wastewater compounds in wastewater effluent and the Big Sioux River in the Upper Big Sioux River Basin, South Dakota, 2003–2004: U.S. Geological Survey Scientific Investigations Report 2005–5249, 117 p.

Snyder, L.R., Kirkland, J.J., and Glajch, J.L., 1997, Practical HPLC method development (2d ed.): New York, Wiley-Interscience, 765 p.

U.S. Environmental Protection Agency, 2005, Guidelines establishing test procedures for the analysis of pollutants (App. B, Part 136, Definition and procedures for the determination of the method detection limit): U.S. Code of Federal Regulations, Title 40, revised as of July 1, 2005, p. 319–322.

Wilde, F.D., Radtke, D.B., Gibs, Jacob, and Iwatsubo, R.T., eds., April 2004, Processing of water samples (version 2.2): U.S. Geological Survey Techniques of Water-Resources Investigations, book 9, chap. A5, accessed January 3, 2008, at *http://water.usgs.gov/owq/FieldManual/chapter5/html/Ch5_contents.html*

Wilkison, D.H., Armstrong, D.J., Norman, R.D., Poulton, B.C., Furlong, E.T., and Zaugg, S.D., 2006, Water quality in the Blue River Basin, Kansas City metropolitan area, Missouri and Kansas, July 1998 to October 2004: U.S. Geological Survey Scientific Investigations Report 2006–5147, 170 p.

Zaugg, S.D., Smith, S.G., Schroeder, M.P., Barber, L.B., and Burkhardt, M.R., 2002, Methods of analysis by the U.S. Geological Survey National Water Quality Laboratory—Determination of wastewater compounds by polystyrene-divinylbenzene solid-phase extraction and capillary-column gas chromatography/mass spectrometry: U.S. Geological Survey Water-Resources Investigations Report 01–4186, 37 p.

Appendixes

Appendix A.   Pharmaceutical Analysis Preparation Data Sheet

Set Position # _____

National Water Quality Laboratory
Lab Schedule 2080 – Solid-Phase Extraction
☐   4200

Lab ID: _____ **Set #:** _____ **Date Received:** _____

- SPE Cartridge Lot No.: _____
    Condition Cartridge: 5 mL MeOH, Vacuum off residual
                                                MeOH, 5 mL H$_2$O

- Sample Weights:
    Bottle + Sample: _____ grams

- Surrogate:
        Solution ID: _____ (5 ng/µL)
    Volume Added: 100 µL

- Spike (QA samples only)
        Solution ID: _____ (2.5 ng/µL)
    Volume Added: 100 µL
                                    _____

- Post-Extraction Weights:
  Bottle + remaining sample: _____ grams

            Empty Bottle + cap: _____ grams   Extraction Date: _____

- Dry Cartridges at 400 mbar until appearance of dry band below top frit, 10 min.

- Test Tube weights: Empty _____ g   Full _____ g

- Elution: Elution Solvent: 6 mL MeOH
                                                4 mL MeOH, 0.09% TFA   Elution Date: _____

- Concentration to approximately 100 µL:
        TurboVap pressure: 5 psi
        Water temperature: 40°C
                        Time: _____

- Reconstitute: Add formate buffer to bring volume up to 1 mL.
- Syringe filter: 0.2 µm PTFE Acrodisc and vial.
- Comments:

Appendix B.   Sample Holding-Time Study

The stability of method pharmaceuticals during storage, either as water samples or in sample extracts, was assessed to determine appropriate holding times for samples prior to extraction and for extracts prior to analysis.

A sample holding-time study was performed assessing the recovery of method pharmaceuticals in fortified filtered water samples held at about 4°C. Organic-free, reagent-water samples, surface-water samples from the South Platte River at Denver, and water samples from the domestic ground-water supply used in recovery studies were filtered in sufficient volume and arranged so as to provide for three 1-L samples of each type to be prepared for analysis on six separate dates over 4 weeks, starting on day 1, with samples extracted and analyzed on days 3, 6, 10, 15, and 28. All samples were filtered, placed into bottles, and fortified simultaneously with 1.00 µg/L (the same solution used for LRS samples), a process that required one full day. Then, three samples of each matrix type were extracted on day 1, and the remaining samples were refrigerated at 4°C until the specified day for their preparation. For the purposes of this study, samples were grouped into sets of nine samples, three of each matrix, and including LRB and LRS samples prepared using organic-free reagent water. Sample extracts were analyzed as soon as possible after completion of the extraction procedure to minimize losses that might occur in the extract and which could not be distinguished from the effects of storage time.

In figures 2 through 5, the mean concentrations of analyses of triplicate samples, fortified at 1.00 µg/L, are plotted in relation to the number of days of storage (at 4°C), ranging from 0 to 28, that each sample set was held. The experiment was conducted for all three matrices. A first-order exponential decay curve was calculated for each pharmaceutical and plotted using the curve-fitting routine provided in the scientific graphing and analysis software used to make the plots (Origin 7.0, OriginLab Corporation, Northampton, Mass). The curve was fitted to each data set using the formula

$$y = y_0 + A_1 e^{-x/t} \qquad (8)$$

where

y = the fitted mean concentration at time x;

x = time, in days;

y_0 = the y offset, an approximate fixed number close to the asymptotic value of the y variable for large values of x;

and

t = the decay constant.

Solving this equation for x equals 0, the curve-fitted concentration at time equals 0 can be obtained. The half-life of each pharmaceutical in each matrix then can be estimated by rearranging the equation and solving for x when y equals the concentration midway between y_0 and y when t equals 0; that is, the time when half the pharmaceuticals between the t equals 0 concentration and the projected final concentration have disappeared. The y-offset concentration (y_0), the curve-fitted concentration on day zero, and the half-life for each pharmaceutical in each matrix are listed in table 29. In one instance, for 1,7-dimethylxanthine in ground water, curve fitting projected a negative y-offset concentration, which resulted in a half-life of 2,490 days. This result was discarded as an artifact of the curve-fitting procedure.

The results in table 29 show that there is substantial variation among pharmaceuticals and among matrices for all three curve-fitting parameters. With the exception of 1,7-dimethylxanthine, the y-offset concentrations suggest that no pharmaceutical concentration in any matrix decays to zero, based upon the 28-day duration of this experiment. The y-offset concentration ranges between 0.160 µg/L for sulfamethoxazole and 0.592 µg/L in surface water, with y-offset concentrations for other matrices falling between these concentrations. Similarly, the curve-fitted concentration on day zero is variable, ranging between 0.178 µg/L for thiabendazole in surface water and 0.900 µg/L for dehydronifedipine in ground water. The calculated half-lives also vary considerably, ranging between 0.295 days for diltiazem in surface water and 42.6 days for cotinine in ground water. The variation observed in the curve-fitting parameters likely results from the same matrix effects previously discussed in this report, that is, matrix effects on compound recovery as observed for sulfamethoxazole and thiabendazole, and matrix enhancement of recovery as observed for caffeine and albuterol, although in this experiment, measured concentrations were all below the fortified concentration of 1.00 µg/L used for all samples.

The half-lives of the pharmaceuticals studied are used as a proxy to estimate appropriate holding times for samples. Given the substantial variation observed in the curve-fitted parameters in table 29, nonparametric statistics, such as the median, the 25th percentile, and the 75th percentile, are used to describe the half-life distribution and to arrive at a single recommended holding time. The median describes central tendency of the distribution, while the 25th and 75th percentiles describe the spread of the distribution. The median half-lives for pharmaceuticals were 3.84, 2.92, and 5.02 days for reagent-, ground-, and surface-water samples, respectively. The 25th percentile half-lives were 3.08, 2.43, and 3.27 days for reagent-, ground-, and surface-water samples, respectively, while the 75th percentile half-lives were 5.14, 6.52, and 9.91 days for reagent-, ground-,

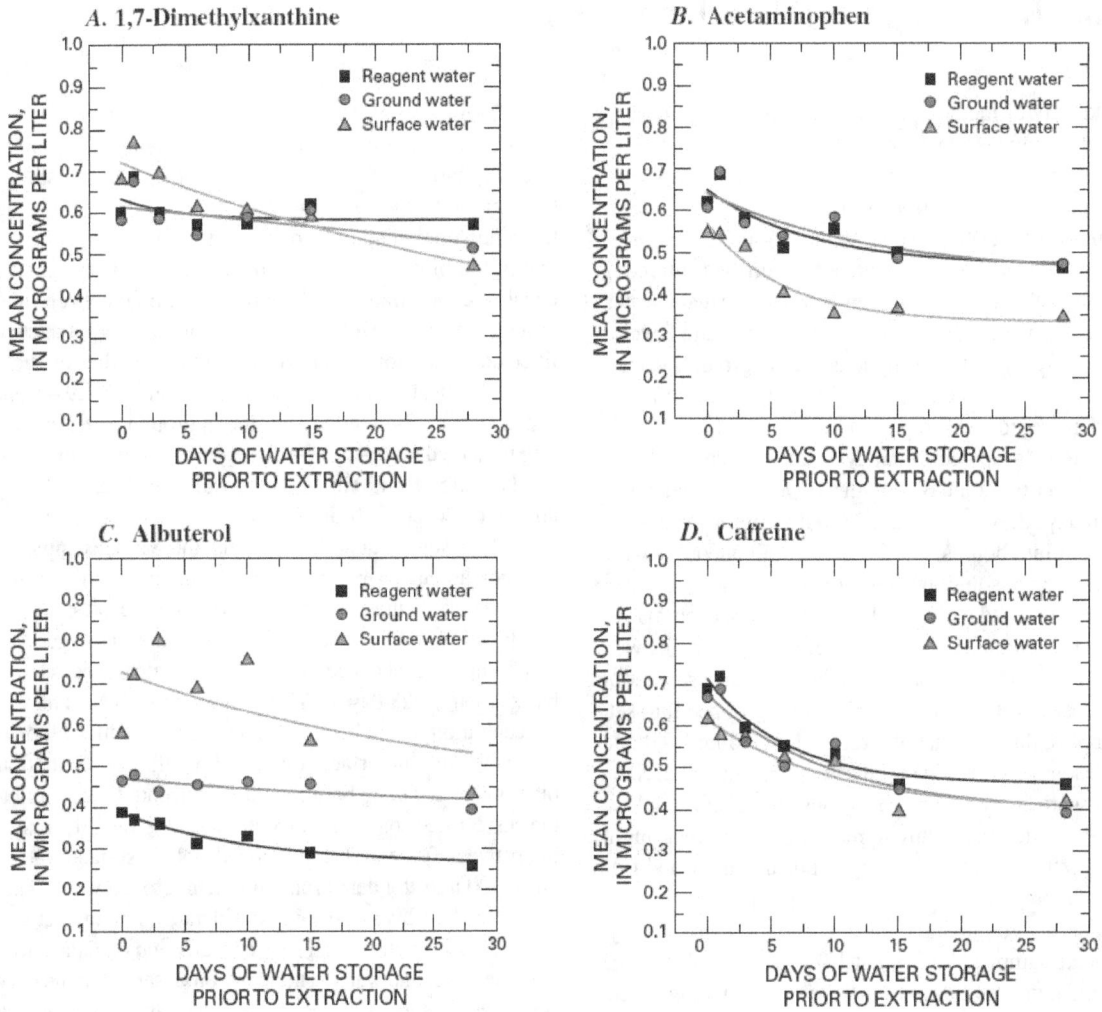

Figure 2. Calculated exponential decay curves for mean concentrations of individual pharmaceuticals from three separate fortified sample-water matrices held at 4 degrees Celsius: (*A*) 1,7-dimethylxanthine, (*B*) acetaminophen, (*C*) albuterol (salbutamol), and (*D*) caffeine.

and surface-water samples, respectively. The overall median half-life, 25th percentile of half-life, and 75th pecentile of half-life were calculated from the data in table 29 and were 4.0, 2.5, and 6.5 days, respectively.

Based on this analysis of half-lives, samples should be extracted no later than 4 days after sample collection to ensure that the analyzed sample is representative of the water sampled at the time of collection. The concentrations of some compounds, such as diltiazem, may be substantially decreased even within this 4-day period; for studies focused on these compounds, even shorter holding times may be necessary. USGS protocols recommend shipping samples by overnight express, and NWQL protocols require that samples are extracted within 48 hours of receipt, so that pharmaceutical samples typically are processed at the NWQL within 3 days of collection.

In using the calculated half-lives to determine a recommended holding time, it is important to recognize that the calculated half-lives estimate the time it takes for a pharmaceutical concentration to decrease between the curve-fitted concentration on day zero to the (non-zero) y-offset concentration, which represents the effective end of first-order decay, based on the 28-day duration of the experiments. The overall median y-offset concentration and curve-fitted concentration on day zero, in micrograms per liter, calculated from the data in table 29, were 0.39 and 0.61 µg/L, respectively. From these medians, the overall median concentration can be calculated at the median half-life, and is 0.50 µg/L (the concentration midway between the overall median y-offset concentration and overall median curve-fitted concentration on day zero), suggesting that, even as it integrates considerable variation

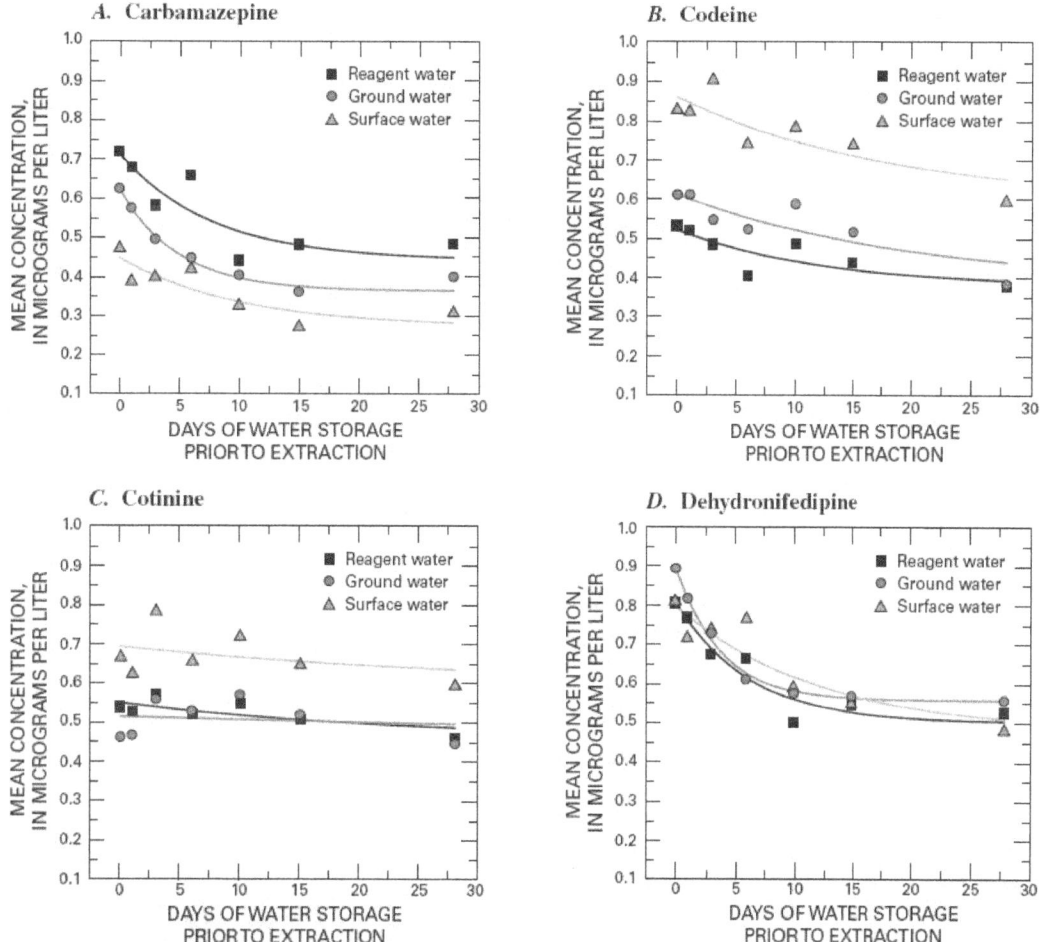

Figure 3. Calculated exponential decay curves for mean concentrations of individual pharmaceuticals from three separate fortified sample-water matrices held at 4 degrees Celsius: (*A*) carbamazepine, (*B*) codeine, (*C*) cotinine, and (*D*) dehydronifedipine.

of the individual pharmaceuticals and the three matrices, the overall median half-life is an appropriate proxy for estimating a holding time.

Similarly, the extracts of the samples prepared on the date of fortification (day 0) of the sample holding-time study discussed above were re-analyzed on seven additional dates to assess losses of pharmaceuticals from sample extracts that were stored over a period that represents the likely maximum range of extract storage. Sample extracts were analyzed by HPLC/MS starting on day 1 and were reanalyzed after 6, 10, 15, 29, 44, 61, and 90 days of refrigerated (4°C) storage. Samples were stored at 4°C because sample extracts awaiting analysis on the HPLC/MS system may be at this temperature for 2 weeks or longer. In figures 6 through 9, the mean concentrations of analyses of triplicate sample extracts, fortified at 1.00 µg/L, are plotted in relation to the number of days of storage at 4°C, ranging from 0 to 90, that each extract was held. The experiment was conducted for all three matrices. A first-order

exponential decay curve was calculated for each pharmaceutical and plotted using the curve-fitting routine provided in the scientific graphing and analysis software used to make the plots (Origin 7.0, OriginLab Corporation, Northampton, Mass.). An analysis of the curve-fitting parameters, similar to that conducted for the water samples, was conducted for the extracts, and is listed in table 30. Curve fitting of the extract holding-time study data was more difficult, as reflected by the greater instances of curves that projected negative *y*-offset concentrations that resulted in unrealistic half-lives, and likely reflects a limitation of applying the curve-fitting algorithim to these data, which, for some compounds, display substantial variation from pure first-order decay. Curve-fitting parameters for dehydronifedipine and sulfamethoxazole in reagent-water extracts, 1,7-dimethylxanthine, acetominophen, codeine, and dehydronifedipine in ground-water extracts, and caffeine, codeine, and dehydronifedepine in surface-water extracts were discarded as artifacts of the curve-fitting algorithm.

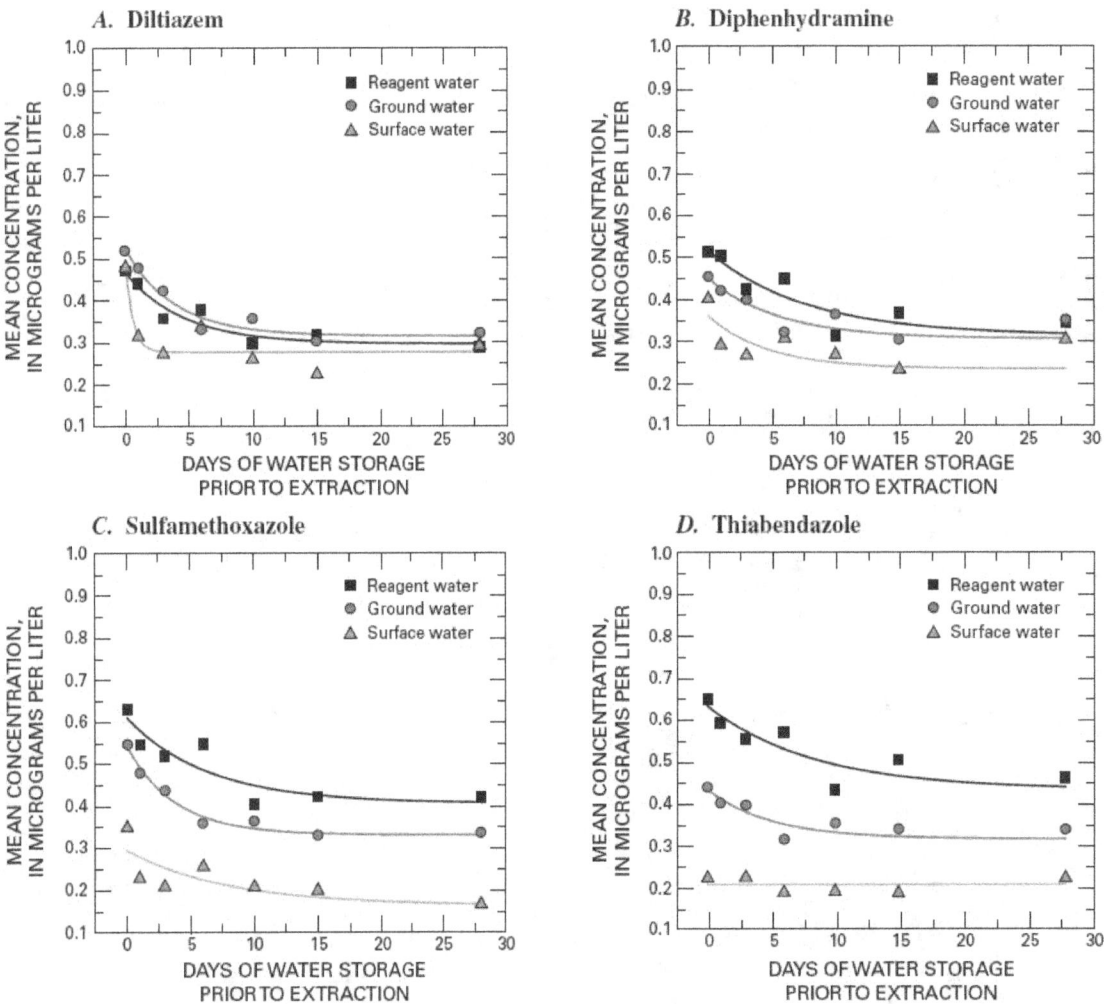

Figure 4. Calculated exponential decay curves for mean concentrations of individual pharmaceuticals from three separate fortified sample-water matrices held at 4 degrees Celsius: (*A*) diltiazem, (*B*) diphenhydramine, (*C*) sulfamethoxazole, and (*D*) thiabendazole.

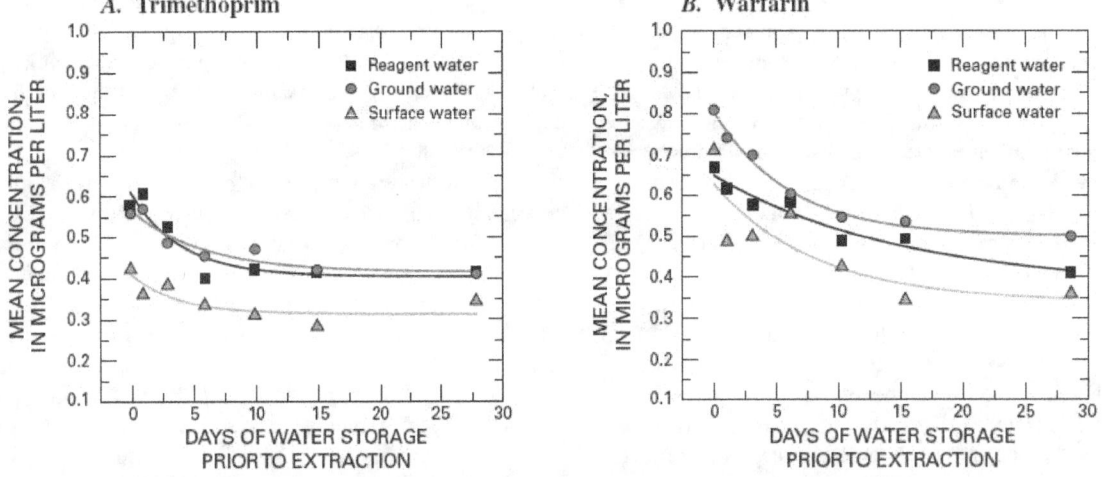

Figure 5. Calculated exponential decay curves for mean concentrations of individual pharmaceuticals from three separate fortified sample-water matrices held at 4 degrees Celsius: (*A*) trimethoprim and (*B*) warfarin.

Table 29.   Curve-fitting parameters and statistical summaries for first-order exponential decay curves applied to the mean concentrations of individual pharmaceuticals from three separate fortified water-sample matrices held at 4 degrees Celsius for 28 days.

[NA, not applicable]

Compound	Reagent water			Ground water			Surface water		
	Y-offset concentration, in micrograms per liter	Curve-fitted concentration on day zero, in micrograms per liter	Half-life, in days	Y-offset concentration, in micrograms per liter	Curve-fitted con-centration on day zero, in micrograms per liter	Half-life, in days	Y-offset concentration, in micrograms per liter	Curve-fitted concentration on day zero, in micrograms per liter	Half-life, in days
1,7-Dimethylxanthine	0.582	0.633	2.29	NA	NA	NA	0.267	0.720	18.09
Acetaminophen	0.465	0.648	4.17	0.445	0.641	6.52	0.330	0.567	3.04
Albuterol	0.252	0.377	5.47	0.386	0.462	12.6	0.422	0.719	12.97
Caffeine	0.457	0.712	3.14	0.389	0.675	4.85	0.391	0.613	5.12
Carbamazepine	0.435	0.711	3.80	0.356	0.619	2.43	0.262	0.441	4.93
Codeine	0.377	0.520	6.19	0.381	0.609	10.1	0.591	0.859	9.19
Cotinine	0.457	0.547	11.1	0.442	0.511	42.6	0.592	0.690	15.4
Dehydronifedipine	0.503	0.811	3.06	0.560	0.900	1.90	0.482	0.801	5.98
Diltiazem	0.296	0.467	2.29	0.315	0.523	1.95	0.277	0.481	0.295
Diphenhydramine	0.311	0.514	3.87	0.303	0.448	2.80	0.232	0.358	2.35
Sulfamethoxazole	0.403	0.606	3.24	0.329	0.538	1.92	0.160	0.291	3.96
Thiabendazole	0.430	0.630	4.17	0.314	0.431	2.49	0.220	0.178	10.14
Trimethoprim	0.404	0.611	2.04	0.415	0.568	2.92	0.314	0.412	1.88
Warfarin	0.373	0.646	7.47	0.499	0.800	2.96	0.339	0.625	3.97
Mean	0.410	0.602	4.45	0.395	0.594	7.39	0.348	0.554	6.95
Median	0.417	0.620	3.84	0.386	0.568	2.92	0.322	0.590	5.02
Standard deviation	0.087	0.111	2.46	0.077	0.137	11.1	0.133	0.201	5.41
Minimum	0.252	0.377	2.04	0.303	0.431	1.90	0.160	0.178	0.30
Maximum	0.582	0.811	11.1	0.560	0.900	42.6	0.592	0.859	18.1
25th Percentile	0.374	0.527	3.08	0.329	0.511	2.43	0.263	0.419	3.27
75th Percentile	0.457	0.648	5.14	0.442	0.641	6.52	0.414	0.712	9.91

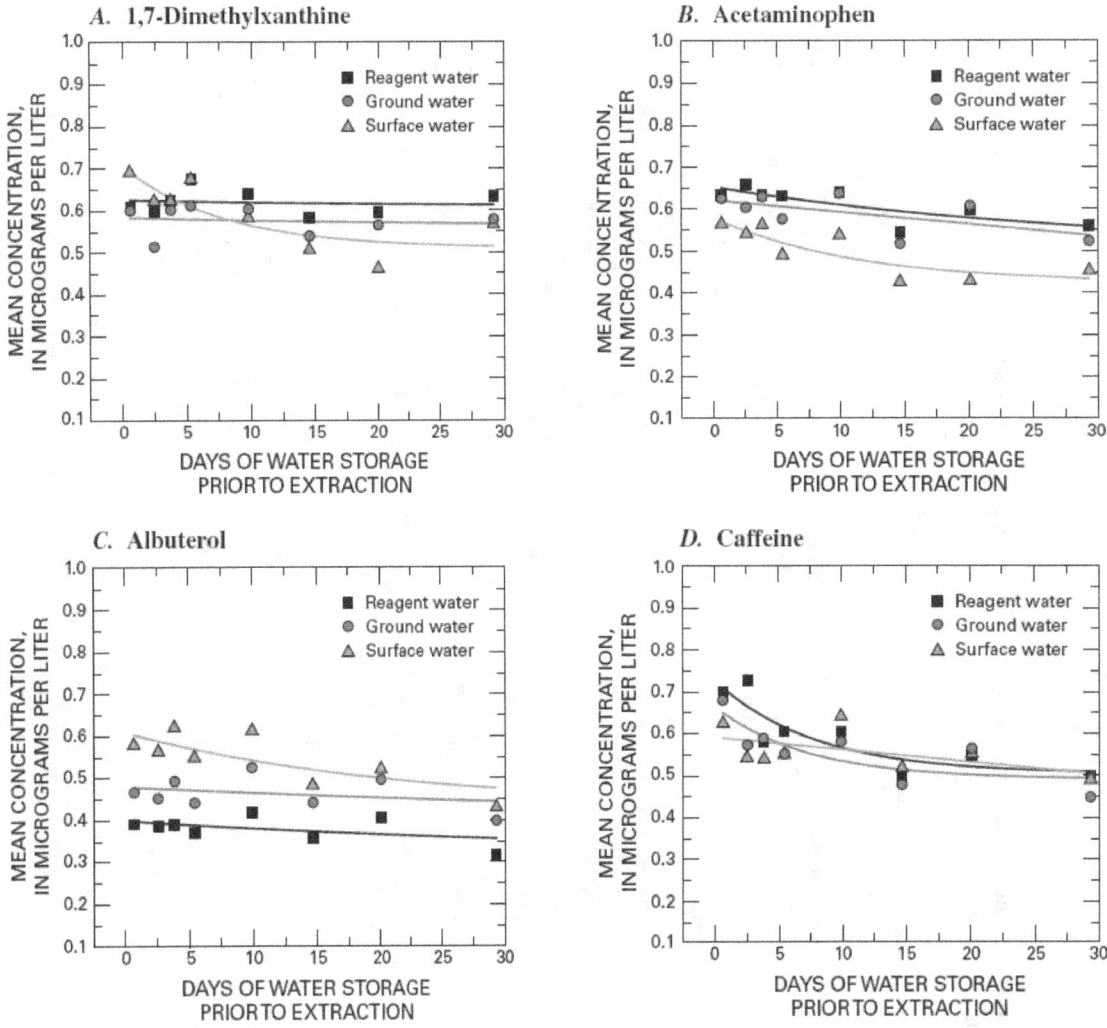

Figure 6.   Calculated exponential decay curves for mean concentrations of individual pharmaceuticals from three stored sample extracts derived from separate fortified sample-water matrices held at 4 degrees Celsius: (*A*) 1,7-dimethylxanthine, (*B*) acetaminophen, (*C*) albuterol (salbutamol), and (*D*) caffeine.

Absolute recoveries of some individual pharmaceuticals, including acetominophen, carbamazepine, sulfamethoxazole, thiabendazole, trimethoprim, and warfarin, were higher for reagent-water and ground-water extracts than for surface-water extracts, for the duration of the experiment. Conversely, recoveries of codeine and cotinine were consistently higher in surface-water extracts than in reagent-water or ground-water extracts. As previously discussed, these differences likely reflect the pharmaceutical-specific effects sample matrix can have upon recovery from extraction and on the suppression or enhancement of ionization. There were distinct differences for individual pharmaceuticals. For example, acetaminophen, cotinine, and dehydronifedipine concentrations (figs. 6 and 7) appeared to decrease at a consistent rate in all matrices, whereas thiabendazole (fig. 8) decreased somewhat in reagent water, but in ground water and surface water remained essentially constant over the 90-day duration of the experiment. Some

pharmaceuticals, most notably dehydronifedipine, appeared to increase in concentration at some point during the experiment, decreasing again at the end of the experiment.

The results in table 30 show that for the reanalyzed sample extracts, there is substantial variation between pharmaceuticals and between matrices for all three curve-fitting parameters. With the exception of dehydronifedipine and sulfamethoxazole in reagent-water extracts, 1,7-dimethylxanthine, acetominophen, codeine, and dehydronifedipine in ground-water extracts, and caffeine, codeine, and dehydronifedepine in surface-water extracts (8 of 42 cases), the y-offset concentrations suggest that no pharmaceutical concentration in any matrix decays to zero, based upon the 90-day duration of this experiment. The y-offset concentration ranges between 0.170 µg/L for sulfamethoxazole in ground water and 0.554 µg/L for 1,7-dimethylxanthine in reagent water, with y-offset concentrations for other matrices falling between these concentrations. Similarly, the curve-fitted concentration on day zero is variable, ranging

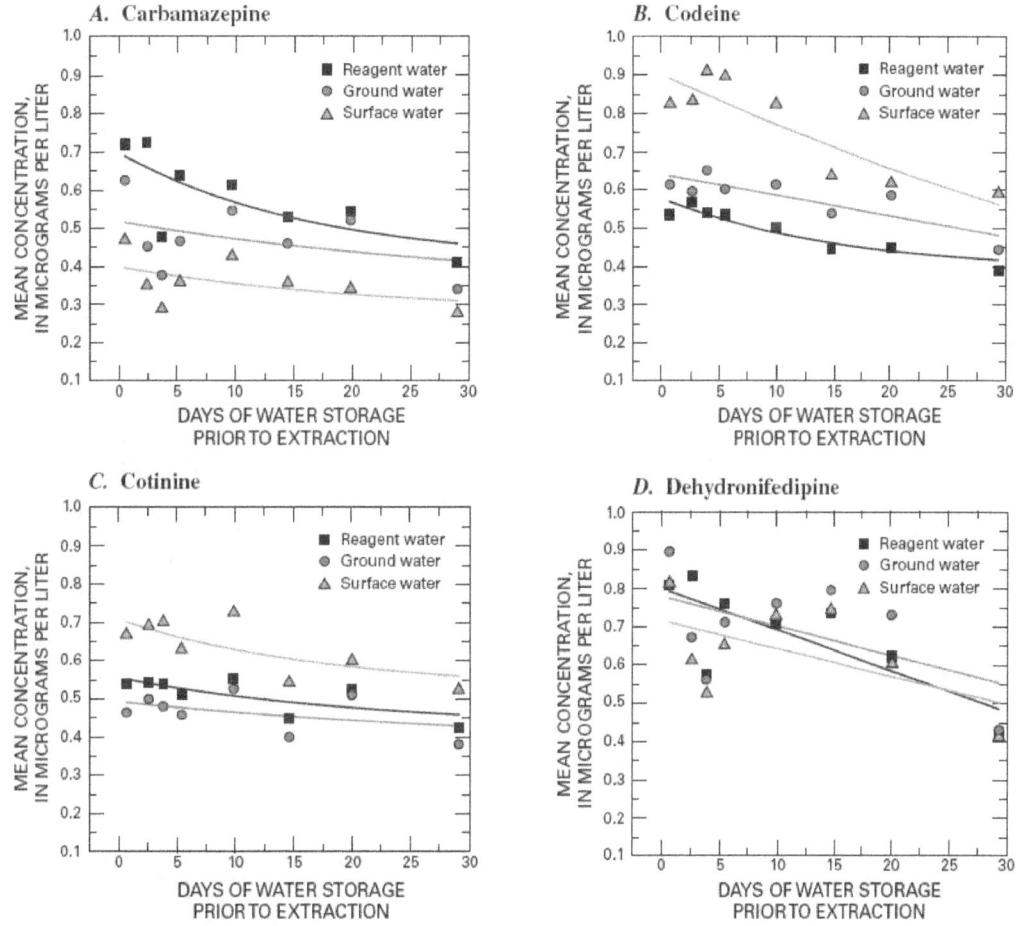

Figure 7.   Calculated exponential decay curves for mean concentrations of individual pharmaceuticals from three stored sample extracts derived from separate fortified sample-water matrices held at 4 degrees Celsius: (*A*) carbamazepine, (*B*) codeine, (*C*) cotinine, and (*D*) dehydronifedipine.

between 0.377 µg/L for albuterol and 0.712 µg/L for caffeine in reagent water. The calculated half-lives range between 6.6 days for 1,7-dimethylxanthine in reagent water and 358 days for thiabendazole in ground water. The variation observed in the curve-fitting parameters likely results from scatter in the data that can be attributed to some of the same matrix effects previously discussed in this report; that is, matrix effects on absolute compound recovery as observed for sulfamethoxazole and thiabendazole, and matrix suppression or enhancement of recovery, although as noted for the water sample holding-time experiment results, measured concentrations were all less than the fortified concentration of 1.00 µg/L used for all samples.

The half-lives of the pharmaceuticals studied are used as a proxy to estimate appropriate holding times for sample extracts. Given the substantial variation observed in the curve-fitting parameters in table 30, the same nonparametric tests applied to the water sample holding-time results, that is the median, the 25th percentile, and the 75th percentile of the distribution, were applied to the sample extract holding-time results in table 30 to characterize the half-life distribution of

sample extracts and to arrive at a single recommended extract holding time. The median half-lives for pharmaceuticals in sample extracts were 26.9, 63.2, and 35.1 days for reagent-, ground-, and surface-water samples, respectively. The 25th percentile of half-lives were 22.4, 53.5, and 26.4 days for reagent-, ground-, and surface-water samples, respectively, while the 75th percentile distribution of half-lives were 36.6, 102, and 43.1 days for reagent-, ground-, and surface-water samples, respectively. The overall median half-life, 25th percentile of half-life, and 75th percentile of half-life for all matrices were calculated from the data in table 29 and were 35, 26, and 59 days, respectively. These results suggest that analysis of pharmaceutical concentrations in extracts produced using this method are acceptable after storage for up to 35 days at 4°C. This finding indicates that assembling analytical batches from multiple sample sets and holding them in the instrument autosampler for 2 weeks or more at 4°C should have minimal effects. Storage at –15°C, the temperature at which most laboratory freezers are operated, would be expected to further minimize storage effects, and is recommended for long-term archival storage of analyzed extracts.

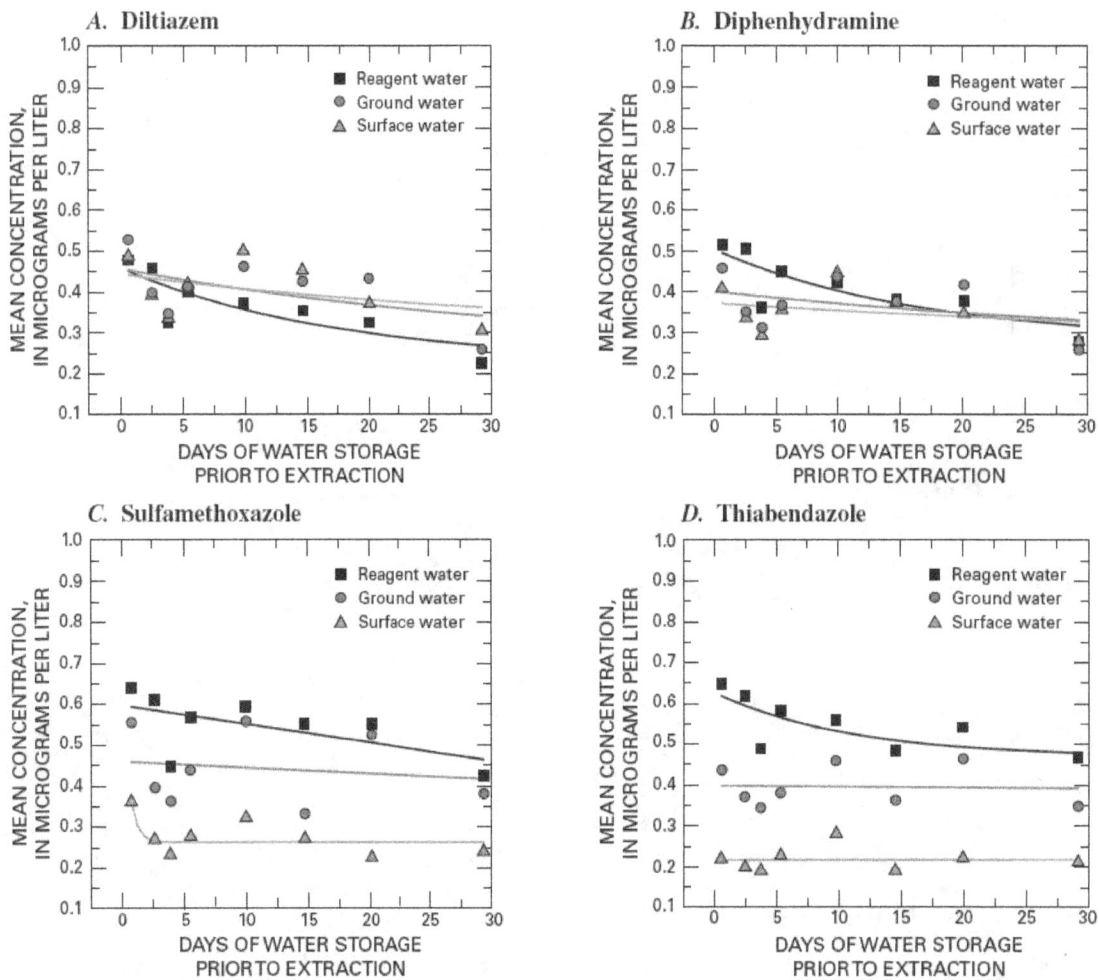

Figure 8.   Calculated exponential decay curves for mean concentrations of individual pharmaceuticals from three stored sample extracts derived from separate fortified sample-water matrices held at 4 degrees Celsius: (*A*) diltiazem, (*B*) diphenhydramine, (*C*) sulfamethoxazole, and (*D*) thiabendazole.

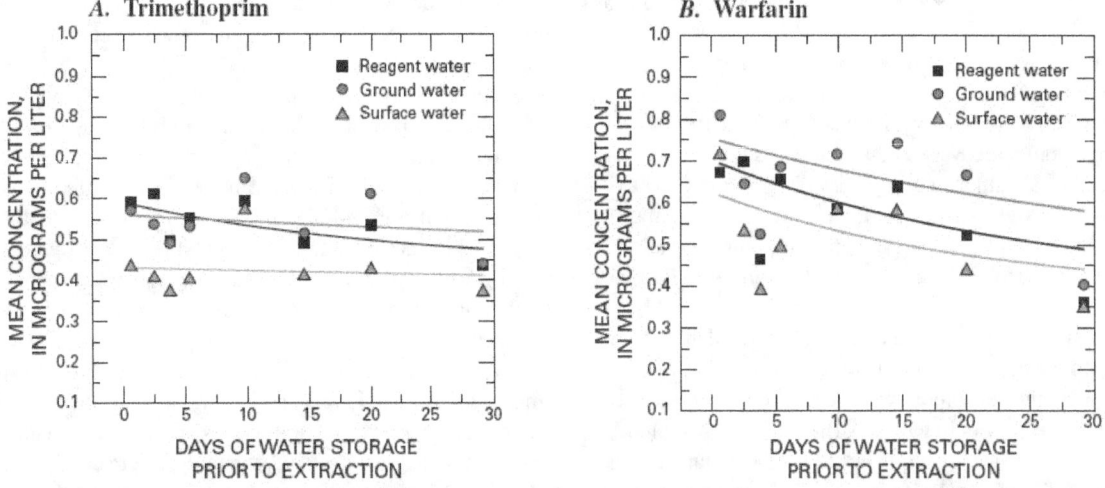

Figure 9.   Calculated exponential decay curves for mean concentrations of individual pharmaceuticals from three stored sample extracts derived from separate fortified sample-water matrices held at 4 degrees Celsius: (*A*) trimethoprim and (*B*) warfarin.

Table 30. Curve-fitting parameters and statistical summaries for first-order exponential decay curves applied to the mean concentrations of individual pharmaceuticals from extracts of three separate fortified water-sample matrices held at 4 degrees Celsius for 90 days.

[NA, not applicable]

Compound	Reagent water			Ground water			Surface water		
	Y-offset concentration, in micrograms per liter	Curve-fitted concentration on day zero, in micrograms per liter	Half-life, in days	Y-offset concentration, in micrograms per liter	Curve-fitted concentration on day zero, in micrograms per liter	Half-life, in days	Y-offset concentration, in micrograms per liter	Curve-fitted concentration on day zero, in micrograms per liter	Half-life, in days
1,7-Dimethylxanthine	0.554	0.634	6.6	3.86	0.593	NA	0.354	0.635	6.58
Acetaminophen	0.484	0.648	46.1	NA	NA	NA	0.401	0.635	46.1
Albuterol	0.303	0.378	62.9	0.387	0.468	79.8	0.420	0.386	62.9
Caffeine	0.488	0.712	12.1	0.474	0.636	11.2	NA	NA	NA
Carbamazepine	0.408	0.711	26.0	0.338	0.512	52.8	0.274	0.687	26.0
Codeine	0.382	0.520	24.1	NA	NA	NA	NA	NA	NA
Cotinine	0.416	0.548	35.4	0.374	0.486	54.4	0.519	0.547	35.4
Dehydronifedipine	NA	NA	NA	NA	NA	NA	NA	NA	NA
Diltiazem	0.214	0.467	27.0	0.246	0.445	53.2	0.294	0.442	27.0
Diphenhydramine	0.269	0.514	26.7	0.250	0.393	67.6	0.270	0.490	26.7
Sulfamethoxazole	NA	NA	NA	0.170	0.446	272	0.248	0.583	12,900
Thiabendazole	0.468	0.630	17.1	0.345	0.399	358	0.279	0.619	17.1
Trimethoprim	0.423	0.611	35.1	0.430	0.548	110	0.356	0.576	35.1
Warfarin	0.349	0.646	40.1	0.394	0.712	58.9	0.335	0.657	40.1
Mean	0.396	0.585	29.9	0.661	0.512	112	0.341	0.569	1,200
Median	0.412	0.620	26.9	0.374	0.486	63.2	0.335	0.583	35.1
Standard deviation	0.0989	0.101	15.4	1.065	0.101	111.8	0.081	0.0945	3,870
Minimum	0.214	0.378	6.6	0.170	0.393	11.2	0.248	0.386	6.58
Maximum	0.554	0.712	62.9	3.86	0.712	358	0.519	0.686	12,900
25th Percentile	0.337	0.518	22.4	0.294	0.446	53.5	0.277	0.518	26.4
75th Percentile	0.472	0.647	36.6	0.412	0.571	102	0.379	0.635	43.1

www.ingramcontent.com/pod-product-compliance
Lightning Source LLC
Chambersburg PA
CBHW081611170526
45166CB00009B/2918